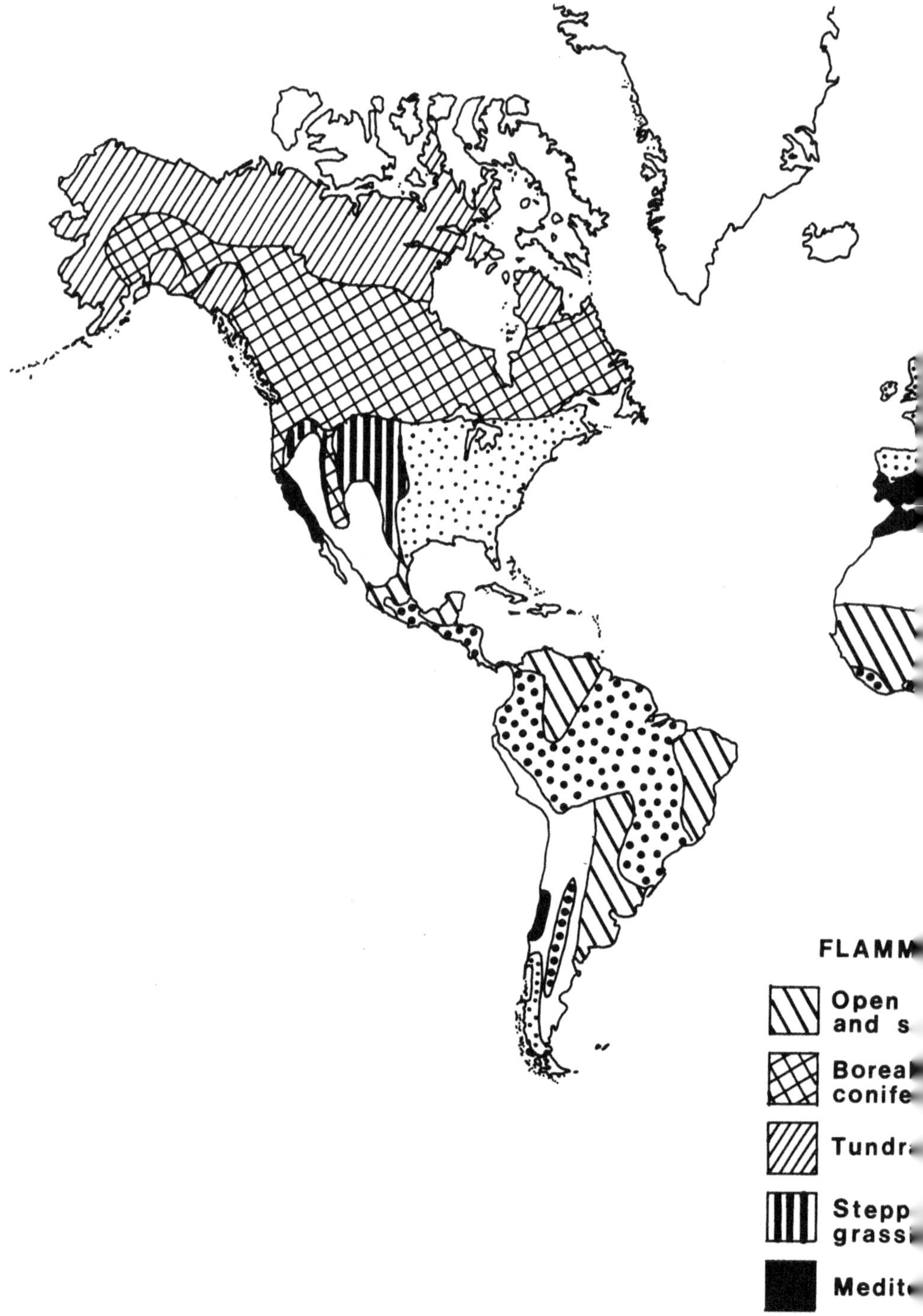

FLAMM
Open
and s
Boreal
conife
Tundr
Stepp
grass
Medit

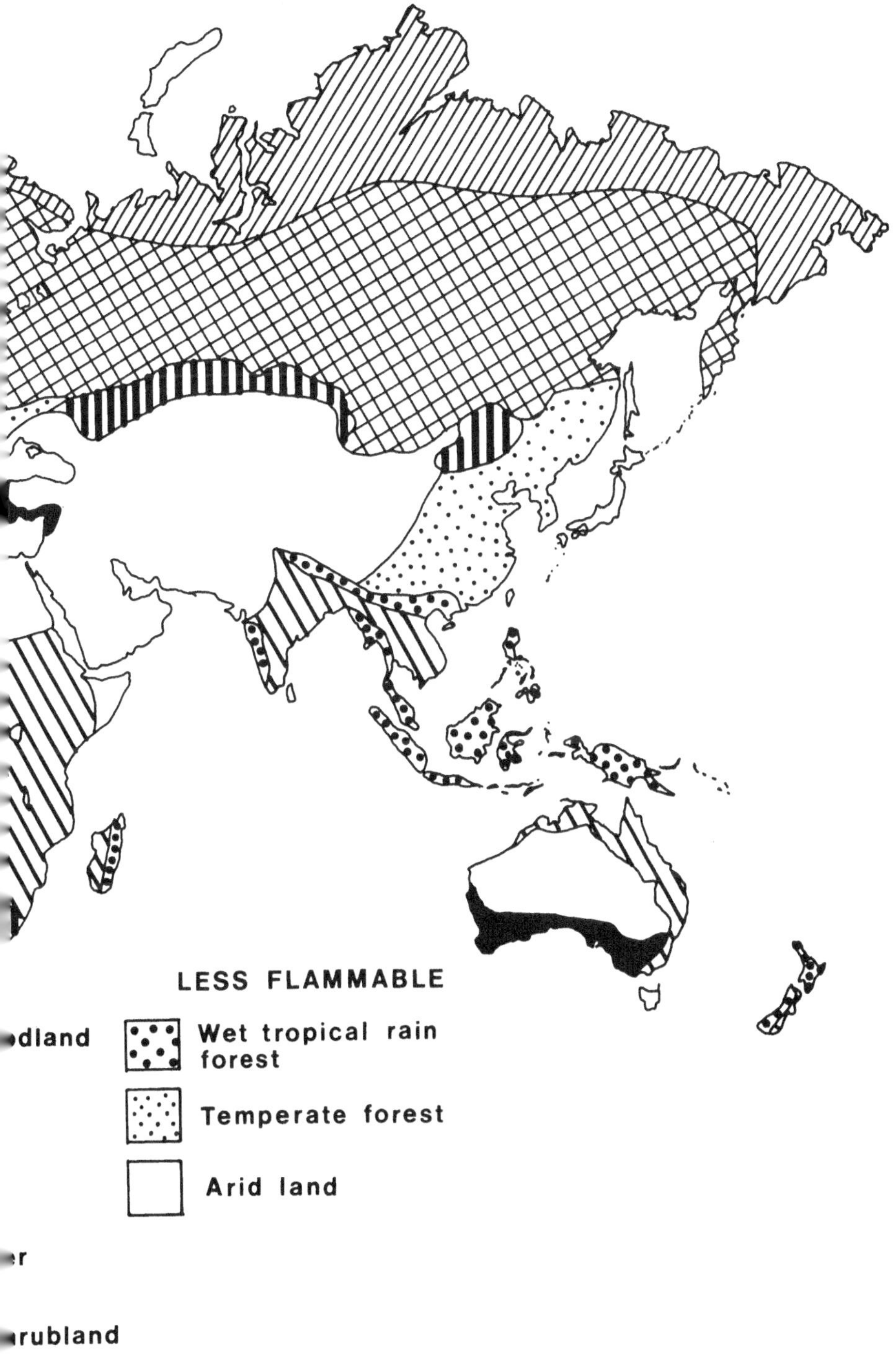

LESS FLAMMABLE
Wet tropical rain forest
Temperate forest
Arid land
dland
er
rubland

FIRE AND PLANTS

Population and Community Biology Series

Principal Editor

Michael B. Usher
Chief Scientific Adviser and Director of Research and Advisory Services, Scottish Natural Heritage, UK

Editors

D.L. DeAngelis
Senior Scientist, Environmental Sciences Division, Oak Ridge National Laboratory, USA

B.F.J. Manly
Director, Centre for Applications of Statistics and Mathematics, University of Otago, New Zealand

The study of both populations and communities is central to the science of ecology. This series of books explores many facets of population biology and the processes that determine the structure and dynamics of communities. Although individual authors are given freedom to develop their subjects in their own way, these books are scientifically rigorous and a quantitative approach to analysing population and community phenomena is often used.

Already published

1. **Population Dynamics of Infectious Diseases: Theory and Applications**
   R.M. Anderson (ed.) (1982) 368pp. Hb.
2. **Food Webs**
   S.L. Pimm (1982) 219pp. Hb/Pb.
3. **Predation**
   R.J. Taylor (1984) 166pp. Hb/Pb.
4. **The Statistics of Natural Selection**
   B.F.J. Manly (1985) 484pp. Hb/Pb.
5. **Multivariate Analysis of Ecological Communities**
   P. Digby & R. Kempton (1987) 206pp. Hb/Pb.
6. **Competition**
   P. Keddy (1989) 202pp. Hb/Pb.
7. **Stage-Structured Populations: Sampling, Analysis and Simulation**
   B.F.J. Manly (1990) 200pp. Hb.
8. **Habitat Structure: The Physical Arrangement of Objects in Space**
   S.S. Bell, E.D. McCoy & H.R. Mushinsky (1991) 452pp. Hb.
9. **Dynamics of Nutrient Cycling and Food Webs**
   D.L. DeAngelis (1992) 285pp. Pb.
10. **Analytical Population Dynamics**
    T. Royama (1992) 387pp. Hb.
11. **Plant Succession: Theory and Prediction**
    D.C. Glenn-Lewin, R.K. Peet & T.T. Veblen (1992) 361pp. Hb.
12. **Risk Assessment in Conservation Biology**
    M.A. Burgman, S. Ferson & R. Akcakaya (1993) 324pp. Hb.
13. **Rarity**
    K.J. Gaston (1994) 224pp. Hb/Pb.

Published by Chapman & Hall, 2–6 Boundary Row, London SE1 8HN, UK

Chapman & Hall, 2–6 Boundary Row, London SE1 8HN, UK

Blackie Academic & Professional, Wester Cleddens Road, Bishopbriggs, Glasgow G64 2NZ, UK

Chapman & Hall GmbH, Pappelallee 3, 69469 Weinheim, Germany

Chapman & Hall USA, 115 Fifth Avenue, New York, NY 10003, USA

Chapman & Hall Japan, ITP-Japan, Kyowa Building, 3F, 2-2-1 Hirakawacho, Chiyoda-ku, Tokyo 102, Japan

Chapman & Hall Australia, 102 Dodds Street, South Melbourne, Victoria 3205, Australia

Chapman & Hall India, R. Seshadri, 32 Second Main Road, CIT East, Madras 600 035, India

First edition 1996

Typeset in 10/12pt Times by Best-set Typesetter Ltd, Hong Kong
Printed in Great Britain by St Edmundsbury Press Ltd, Bury St Edmunds, Suffolk

ISBN 0 412 47540 5

A catalogue record for this book is available from the British Library

Library of Congress Catalog Card Number: 95–71231

Printed on permanent acid-free text paper, manufactured in accordance with ANSI/NISO Z39.48-1992 and ANSI/NISO Z39.48-1984 (Permanence of Paper).

FIRE AND PLANTS

William J. Bond
Department of Botany, University of Cape Town, South Africa

and

Brian W. van Wilgen
Division of Forest Science and Technology, CSIR, Stellenbosch, South Africa

CHAPMAN & HALL
London · Glasgow · Weinheim · New York · Tokyo · Melbourne · Madras

Contents

Preface

In the part of the world where we live, ecology cannot be studied without reference to fire. Nor can ecological principles be applied to ecosystem management without some understanding of the role of fire. Yet when we turn to the general ecological literature for theory or models relevant to these systems, we find them disturbingly mute. Ecology textbooks, the source of the principles and problems that guide the science, barely mention fire despite its importance in this and other parts of the globe. Population modelling approaches developed for non-fire-prone systems are often misleading and have led to wasted effort when applied to fire-dependent species. Community theory developed elsewhere has also led to misconceptions that have taken years to sort out. Fire ecologists have developed their own approaches to population and community ecology. However the strongly applied context has hidden much of their work from more theoretically inclined ecologists.

This book attempts to make the very large literature on fire ecology accessible to a wider readership by describing general patterns, and drawing out common principles and common problems in the diverse vegetation types affected by fire. We have tried to place fire research in a more general context, indicating similarities and differences with more commonly discussed processes in the general literature on population and community ecology. We believe that many areas of ecological theory can be greatly enriched by explicitly including fire, along with related processes such as herbivory, in searching for an understanding of how the natural world works.

On the other hand, fire ecologists have neglected some areas of research that are of great interest to theoretical ecologists. For example, there have been few studies on species coexistence and diversity patterns in communities affected by fire. There is still no general population model that attempts to explain the intrinsically unstable mix of trees and grasses in savannas. There has been very little work on the evolutionary importance of fire, either as a selective pressure producing adaptive responses or as a biogeographic agent moulding the composition and characteristics of vegetation. Throughout this book we have drawn attention to gaps in ecological research on fire and questions of general interest to ecologists for which we have very few answers.

Ultimately we hope to contribute to the integration of fire ecology into general population and community ecology. The exercise is bound to irritate both sides. Fire ecologists, from their applied tradition, may be reluctant to embrace general models or theory. Theoretical ecologists have been reluctant to delve into the applied literature. We obviously believe the effort for both is worthwhile.

As a consequence of avoiding fire in formal ecological training, biologists concerned with conserving biological diversity still display ignorance, and often emotive behaviour, when dealing with fire-related issues. Too often this ignorance of fire has resulted in serious misjudgments in the conservation or management of fire-prone ecosystems. It is still widely seen as an agent destroying the 'balance of nature' rather than a process that helps preserve much of the world's diversity. Similar deficiencies in training have, we believe, led to insufficient attention being given to fire by the new generation of global biogeographers tackling the question of biological responses to global warming.

A book which attempts to straddle the technical side of fire research and conceptual ecology requires a great deal of mental contortion if it is to be written by a single author. Brian van Wilgen wrote most of chapters 2 and 8, dealing with the former, and William Bond most of the remainder. South Africa has produced several fine fire ecologists to whom we owe a considerable debt. We have been variously influenced by John Phillips, pioneering giant of African ecology, Neil Tainton, Fred Kruger, Eugene Moll and Winston Trollope. Dick Yeaton has provided a refreshing alternative perspective. Our colleagues Richard Cowling, Dave Richardson and David Le Maitre read chapters and provided useful comment. Jeremy Midgley read most of the book and it benefited greatly from his keen insights and criticisms. On other continents, we have profited over the years from contact with colleagues in the MEDECOS network, especially: in North America, Martin Cody, Jon Keeley, Hal Mooney, Phil Riggan, Phil Rundel, Paul Zedler; in Australia, Malcolm Gill, Richard Hobbs (ex Scotland), Byron Lamont, Ray Specht; in Europe, Jose Moreno. Other colleagues who have enriched our understanding of the physical phenomenon of fire include Frank Albini, Bob Burgan and Dick Rothermel of the US Forest Service, Brian Stocks from Forestry Canada, and Johann Goldammer from the Max Planck Institute, Germany. Steve Pyne, from Arizona, also opened our eyes to aspects of the cultural history of fire. We have been very fortunate to have Michael Usher as our editor and thank him for his patience and much helpful advice.

Kristal Maze and Sandy Smuts helped in many ways. Thank you both. Ann Green drew most of the figures. The Foundation for Research and Development and the Department of Environment Affairs helped with funding.

Finally WJB thanks Winifred, Elizabeth, Helen and Christopher for putting up with 'the book' so patiently while BVW thanks Jane, Nicola and Lawrence for their tolerance.

1 Introduction

1.1 THE GLOBAL IMPORTANCE OF FIRE

After human urban and agricultural activities, fire is the most ubiquitous terrestrial disturbance. From Arctic tundra and boreal forests to tropical grasslands and savanna, fire annually consumes enormous quantities of plant biomass (see end papers). In the tropics alone, it has been estimated that 2700–6800 million tonnes of plant carbon is burnt annually, mostly in savanna fires or in shifting agriculture (Crutzen and Andreae, 1990). Terrestrial vegetation has been swept by fire since at least Mesozoic times. Since the mid-Tertiary, the lethal combination of grasses and fire helped create open plains and initiate the great radiation of grazing mammals. Hominids first mastered fire over a million years ago (Brain and Sillen, 1988), setting in motion a great wave of environmental transformation which continues today. In Australia, the vegetation of the continent was transformed first by Aboriginal firestick farming and then by the burning practices of European settlers (Singh *et al.*, 1981; Kershaw, 1986; Pyne, 1991). In North America, fire set by the first human settlers expanded the prairies and ate into woodlands and forests (Pyne, 1982; Axelrod, 1985). In Africa, evergreen forests shrank and grasslands and savanna expanded. People, fire and vegetation continue to interact in modern landscapes. The evergreen forests of Hawaii are shrinking due to the spread of non-native grasses and the grassland fires they generate (D'Antonio and Vitousek, 1992). In the tropics, savanna burning continues to erode the boundaries of tropical forests (Stott, 1988; Crutzen and Andreae, 1990; Backeus, 1992). In Madagascar, the unique endemic flora is shrinking under the combined effects of farming and fire (Green and Sussman, 1990). In many industrial countries, the process of forest destruction by fire has been reversed. Public antipathy to fire has led to its suppression, also with profound effects on vegetation patterns. Woodlands have replaced grasslands, forests replaced shrublands, broad-leaved trees invaded conifers. Fire is a very general and influential ecological phenomenon.

1.2 WHY WE HAVE WRITTEN THIS BOOK

Research on fire and its effects has been conducted by foresters, range managers and conservation biologists since the early decades of the twentieth century. There is a large applied literature on the topic, much of it narrowly focused on particular ecosystems or particular applied problems.

Table 1.1 The number of pages indexed for different ecological processes in current textbooks on introductory ecology. Books marked with an asterisk are general textbooks on applied ecology and might be expected to have a relatively greater emphasis on fire

Textbook	*Competition*	*Predation*	*Herbivory*	*Disturbance*	*Fire*
Begon *et al.* (1990)	110	94	54	50	4
Colinvaux (1986)	31	86	15	2	0
Ehrlich and Roughgarden (1987)	47	40	29	12	1
Putman and Wratten (1984)	61	42	11	0	0
Ricklefs (1990)	40	30	7	4	3
Stiling (1992)	43	34	53	0	12
Beeby (1993)*	17	8	0	18	1
Newman (1993)*	2	7	33	9	6
Smith (1990)*	79	35	30	37	15

The applied emphasis of fire ecology seems to have isolated the topic from ecologists of a more theoretical bent. Discussion of fire and its effects is conspicuously lacking in general ecological textbooks despite large sections devoted to competition and predation and smaller sections on related topics such as herbivory and disturbance (Table 1.1). Since these books introduce the conceptual foundation of the science, it would seem that fire ecology has produced no concepts of general interest or examples to illustrate such concepts. Alternatively, any general insights might be buried so deeply in the applied literature that they have escaped notice by ecologists who write textbooks. We believe the latter is more likely. Fire ecology may also have suffered from a philosophical bias. Most textbooks are written in that small part of the globe which does not burn regularly – the temperate deciduous forest belt. This scientifically productive area also produced the powerful concept of the 'balance of nature'. Fire does not fit easily into nature in balance. It has taken many years to replace the European antipathy to fire with an appreciation for its role in areas influenced by Western thought (Pyne (1982, 1991) gives excellent accounts of the cultural history of wildland fire). We believe that fire ecology needs greater emphasis in ecological training. This book is intended to bridge the gap between the large applied literature on fire ecology and the conceptual literature which forms the foundation of the science.

1.3 WHAT THIS BOOK IS ABOUT

This book is an attempt to make the very large applied literature on fire ecology accessible to a wider audience. We do so in the context of the series of which this book is a part – by considering population and community ecology of plants affected by fire. We have not considered the ecology of animals and fire except where they affect plant populations. The effects of fire on ecosystem processes are important and well studied but are also peripheral to our central theme of populations and communities. The book is a synthesis, rather than a review. Instead of cataloguing the many studies in different systems and in different parts of the world, we have emphasized general models, shared patterns of responses and possible mechanisms that may underlie them. This approach carries the risk of oversimplification and broad generalization in an area dominated by empirical studies. However, general models have value, if only as targets for shooting down. Our attempts at synthesis have exposed many gaps. We have tried to point the reader to these as opportunities for further research.

The parochial nature of fire ecology has restricted the spread of theory or general concepts. Communication among ecologists working on similar systems in different continents has generally been weak. This is unfortunate because much that is implicit in research approaches used by fire ecologists is at variance with widely accepted theory. We attempt to restate implicit assumptions as explicit hypotheses or models. For theoreticians, there is much scope for developing verbal models into more formal theory. For the empiricist, some fire-prone systems provide excellent opportunities for experimental testing of current theory that are difficult to match elsewhere. There are also many data sets from long-term burning experiments which can be used to address new questions, such as the influence of disturbance on different scales of community diversity (Collins, 1992).

The books that have been written on fire ecology are often organized around a single major formation or geographic region. We have tried to include studies from all formations where fire is common, although much of our research experience is in Mediterranean-type shrublands. We have also used studies from a variety of geographic regions to illustrate generalities in fire responses. The biology of many of these systems may appear bizarre and unfamiliar to readers from other continents. We hope it acts as an incentive to travel. Much of the fire literature is difficult to obtain outside the country of origin. We have used journals and books most easily available to us. Inevitably our own geographic region, Africa and its south-western tip, home of the Cape Floral Kingdom, features prominently.

1.4 METHODS OF FIRE ECOLOGY

Studies of the ecological effects of fire have incorporated a variety of methods differing with different systems. Replicated field experiments, involving burning at different frequencies or seasons, have been used widely in grasslands and savanna. These include some of the longest-running field experiments in ecology (Rothamsted excepted!). Some of these are poorly replicated and would fail modern criteria for experimental design. Nevertheless they give invaluable indications of the long-term effects of different fire treatments. Where fire return intervals are longer, in the order of decades or centuries, much use has been made of chronosequences, stands of known age since the last burn. Though chronosequences assume that only stand age varies and that plots are otherwise homogeneous, they are often the only practical way of obtaining data for long-lived species. Simulation studies have also been widely used to model vegetation change where fire frequencies are long. Many of the more recent models incorporate spatial effects. Finally, studies of fire scars, pollen and charcoal deposits have extended our understanding of the synergy between vegetation and fire over centuries in the past, and at spatial scales from points to landscapes.

Some widely used approaches in ecology have been comparatively neglected in fire studies. In striking contrast to herbivory, there are few physiological studies of the effect of fire on plants or on 'plant defences' against defoliation by fire. Also, in contrast to many areas of ecology, an evolutionary perspective has been conspicuously lacking. Both areas offer rich opportunities for further study.

Feedback effects between fire and vegetation have mostly been studied indirectly. Research on the physical phenomenon of fire has been driven by a need to control wildfires. The properties of plants that allow them to burn, and determine how and when they will burn, have formed a major focus for this work. These studies have contributed to an understanding of factors governing the interaction between fire and vegetation.

1.5 ECOLOGICAL CONCEPTS AND FIRE ECOLOGY

One reason why fire may have been neglected in general ecology textbooks is uncertainty as to where to place the material. Fire falls somewhat uneasily between physical and biological processes impacting plants. The physical effects of fire fall within the ambit of disturbance ecology. But fire also has much in common with herbivory, especially vertebrate herbivory, and may profitably be viewed as a major generalist herbivore of plant biomass.

The importance of disturbance in the dynamics of plants and animals has been widely recognized in the past decade (Sousa, 1984; Pickett and White, 1985). Disturbance has been defined as 'any relatively discrete event in time that removes organisms and opens up space which can be colonized by individuals of the same or different species' (Begon *et al.*, 1990). Fire is a discrete event 'opening up space' at scales from less than 1 ha to more than 1 million ha at intervals from 1 to 1000 years, depending on the nature of the fuels and the occurrence of conditions conducive to fire.

As a 'herbivore' fire consumes large amounts of plant material in irregular 'bites' of varying intensity and season. There are many similarities between plant responses to herbivores and to fire – and some important differences. Herbivory can influence plant growth, survival and fecundity and can impact the dynamics of seeds and seedlings (Crawley, 1983). So too can fire. However, the effects are often much greater and of different sign. Fire is one of the few perturbations that, alone or in combination with other forces, regularly kills mature plants. This makes it an important agent in structuring communities, since the new openings created provide the potential for vegetation change. Amongst herbivores, perhaps only elephants or bark-feeding vertebrates approach fires in their importance as killing agents of large trees. In contrast, fire is only one factor amongst many causing the death of seedlings. Herbivores usually suppress plant reproduction but the effects of fire are more complex. Many species are stimulated by fire, showing enhanced flowering, seed germination and seedling recruitment. Some have an obligate dependence on fire for reproduction and can be driven to extinction if fires are suppressed.

One of the most general attempts to understand the role of herbivores in determining ecosystem patterns was that of Hairston *et al.*, (1960). They believed that top-down control of plant biomass by herbivores, or of herbivores by carnivores, was a major factor determining ecosystem structure. This seminal paper initiated a large number of studies in marine and freshwater habitats and some in terrestrial environments. The theory of trophic control has been claimed to be the central theory of ecology (Fretwell, 1987) since it includes other ideas of agents important in structuring ecosystems. For example, it specifies where competition between plants will be important and where it should be unimportant in determining vegetation structure (Oksanen *et al.*, 1981; Oksanen, 1988). Fire has not been included in the development of trophic theory. However, as a major consumer of biomass, we suggest that it, too, falls within the ambit of the theory (section 7.6). In a sense, fire competes with herbivores for 'fuel'. Exclusion of fire can cause major changes to plant biomass or have very little effect, depending partly on the importance of herbivores. Certainly, explicit incorporation of fire would strengthen the predictions of trophic theory with regard to the effects of manipulating different trophic levels.

1.6 FIRE AND THE ECOLOGY OF PLANTS

This book has been arranged to take the reader through the spectrum of topics that constitute the subject area of fire and its effects on plants, plant populations and communities. We begin with an account of how and why plants burn. This is followed by sections concentrating on the responses of individual plants to fire, population dynamics, evolutionary ecology, dynamics of plant communities and a brief account of how fire ecology can be used in the management of ecosystems. We conclude with a chapter on the large-scale effects of fire on vegetation and how the distribution of vegetation may change with changing fire regimes and global warming.

The remainder of this chapter is intended to provide an introduction to the topics covered in the book, and should provide the reader with an overall perspective when reading the following chapters.

1.6.1 Determinants of fire

Analytical models have been developed for plant/herbivore interactions but not for fire/plant interactions. The closest is perhaps an attempt by Clark (1988) to model feedbacks between fuel accumulation rate and fire frequency. Clark used plant biomass to represent fuel – the 'food' of fire. The effects of fuel availability on fire were then measured as a change in fire frequency. However, the incidence of fire depends not only on the properties of fuels, the biological component, but also on appropriate weather conditions and ignition sources to initiate and propagate fires. Clark included the weather component as a second independent variable interacting with fuel availability to determine changes in fire frequency.

Most models of fire behaviour were developed for applied purposes. They incorporate physical and empirical functions to simulate fire behaviour (Chapter 2). The models are effective representations of real fires and have been used for guiding fire control operations. They are useful in indicating the relative importance of weather and fuel conditions for spread, and behaviour, of fires (section 2.2). These models indicate that fires are much more likely to occur in some communities than in others, depending on their fuel properties. Fuel properties depend on the physical mixture of species, as well as their individual properties (section 2.3). Fire hazard is thus a collective property of a community of plants. This, too, complicates the development of analytical models analogous to Lotka–Volterra-type herbivore/plant models for exploring feedbacks between fire and vegetation.

1.6.2 Plant responses to burning

The response of vegetation to fire is very variable. Crown fires can destroy ancient conifer forests and initiate a recovery process that may take decades or centuries to complete. In grassy communities recovery may be so rapid that the effects of a burn may disappear within weeks. Population change in some woody plants is dominated by the effects of fire. Others may barely be affected. Population dynamic models help interpret changes in populations. Population models were first developed by animal ecologists. Refinements for plants have included consideration of their sessile nature in models of interactions (Pacala and Silander, 1985) and their modular structure (Harper, 1977). Another way in which population dynamics of plants differs from animals is in the importance of individual persistence. Plant species vary greatly in the capacity for individual persistence. Persistence is not simply a matter of longevity. Some conifers live for thousands of years but can be killed by a single intense fire. Others, especially clonal species, are also long-lived but survive repeated defoliation by fire or herbivores. The capacity for persistence, in the face of repeated burning, varies greatly among species and for different size classes within a species. Some understanding of the capacity for persistence is an essential starting point for fire demography.

Patterns of persistence fall into two broad categories: sprouters and non-sprouters. Sprouters have the capacity to regenerate from buds, even if the canopy is consumed in a crown fire. Persistence depends on insulation of buds from fire heat, for example by thick bark or layers of soil (section 3.1). Meristematic tissue which initiates sprouting is located in different plant parts in different species, resulting in diverse responses to burning. Size-dependent variation in sprouting within a species is an important factor governing fire effects. Many species characteristic of fire-prone vegetation cannot sprout and are killed by crown fires. Others are not killed by fire but die in the interval between fires. However, their seeds persist in the community as dormant seedbanks and they reappear after the next burn. Non-sprouting species include some of the most specialized fire-dependent life histories.

Along with persistence, fire also influences recruitment. Again, two broad categories of fire response can be identified: species whose recruitment is stimulated by fire (we call them fire-recruiters) and species which recruit at other times responding to other cues (non-fire-recruiting species). Many species have a near obligate dependence on fire for recruitment (section 3.3). Gap-phase recruitment has been studied extensively in forests and herbaceous communities. Gaps favour recruitment for a variety of reasons, including increased availability of resources, and reduction of predation and competition. Where fire regularly forms the gap, a host of specialized traits have evolved that cue recruitment to coincide with post-burn conditions. They include fire-stimulated flowering (Plate 1), seed release (Plate 2) and germination. Investigation of these traits is still far from complete. Smoke-stimulated germination was first described for a single fynbos species in 1990 but has since been found in scores of species on three continents (section 3.3.3).

Ecologists have been slow to recognize fire-recruiting cues, partly because of the historical antipathy to fire and the view that it is an unnatural disturbance. The consequences for conserving diversity can be disastrous. To illustrate the point we use an example from the Cape fynbos. The marsh rose, *Orothamnus zeyheri*, is a very rare and exceptionally beautiful shrub member of the Proteaceae (Boucher, 1981). After it was first described in 1848, the location of the species remained a mystery for over 60 years until it was rediscovered in the flower markets of Cape Town. Cut flowers were traced from the flower markets to their source in a small area of rugged mountains in the south-western Cape. Following the advice of the most eminent botanists of the time, populations were carefully protected from fire and the public. However, wild populations had dwindled to a few single plants by the late 1960s. Fortunately, an astute forester noticed seedlings emerging in regularly burnt firebreaks and a cautious programme of prescribed burning was initiated. Hundreds of *Orothamnus* seedlings began emerging. Populations known to have gone extinct were extinct no longer. Seedlings appeared where the species had never been seen before. Today *Orothamnus* is thriving, with its obligate requirement for fire (it is a fire-recruiting non-sprouter) well understood (Boucher, 1981).

Cases such as this indicate the cost of neglecting the basic biology of species when implementing conservation actions. Many thousands of species depend, like *Orothamnus*, on fire for their continued existence.

1.6.3 Fire and plant population dynamics

The central focus of this book is on how fire affects changes in plant populations and communities. We approach the problem by considering three alternative hypotheses for how fire drives population change (Chapters 4 and 5). The interval-dependent hypothesis argues that the length of the fire interval is of central importance. It is closest to the standard body of population theory and employs standard time-dependent population models. Growth, mortality and fecundity determine population growth. For example, population growth in a species killed by fire would depend on the number of seeds available at different intervals since the last burn. This would be calculated from survivorship and fecundity schedules coupled with seedbank dynamics. Models of this kind have been developed for a number of species using data collected from chronosequences, and cohort analyses. We review a variety of models used to simulate population growth and the effects of varying fire interval based on this information (section 4.3).

Most of these models assume unlimited resources and exponential growth. Density dependence has often been ignored in fire demography. Survivorship and fecundity are typically represented as functions of time since last fire and not of neighbourhood effects (cf. Pacala and Silander, 1985). However, at least in theory, density dependence can lead to endogenous population oscillations. These can occur even where fire frequencies and other environmental conditions are held constant (May, 1974). Under some conditions, density-dependent feedbacks cause growth trajectories resembling a random walk, appropriately known as chaos. Plants may be unlikely candidates for chaotic dynamics. Density-dependent thinning, which is widespread in plant populations, dampens feedback effects (section 4.5). Fecundity often reaches an asymptote with increasing density, also damping feedback effects of density on regeneration. There are very few plant examples of population fluctuations attributed to density dependence, and these are controversial. Nevertheless, simple logistic models of some non-sprouting shrub species suggest that oscillations observed in nature may be of endogenous origin (section 4.5.3). Perennial grasses may also show endogenous oscillations because of density-dependent feedback between biomass production and litter accumulation (Tilman and Wedin, 1991). Where neither fire nor herbivores remove surplus biomass, these species would undergo self-induced cycles of low and high biomass (section 4.5.4). These studies suggest that density-driven oscillations should not be dismissed too quickly as explanations for population fluctuations. They also suggest that far more attention should be given to density dependence in fire demography studies.

The third class of explanations for population fluctuations concerns the effects of particular fires. We have called these effects 'event-dependent'. Event-dependent population regulation is important in most fire-prone populations. Fires vary in their season, intensity, the size of the burnt area and the weather conditions after the burn. Experimental and descriptive studies have shown that variation in the fire event can completely override interval-dependent effects on recovery (Chapter 5). For example, savannas burnt at the same interval can be transformed to open parklands or dense woodlands depending on the season of burn.

Consideration of both the 'regeneration niche' (Grubb, 1977) and the 'persistence niche' are central for predicting event-dependent effects. Both models and mechanics driving event-dependent effects fall outside the ambit of conventional population ecology (sections 5.8 and 5.9). Important factors governing persistence and regeneration include the location and dynamics of regenerative tissue, both buds and seeds, for interpreting fire intensity effects. Phenology of growth, carbon and seed reserves are important for understanding fire season effects. Several models of varying generality have been proposed to explain event-dependent effects. We have distilled them into a list of key attributes which determine population responses to fire at different seasons and of different intensities (section 5.10). The attributes allow qualitative prediction of the effects of fire on different species. They are quite different from, but analogous to, widely known systems that rank species according to their competitive ability, resource use and regeneration requirements (e.g. Grime, 1979; Tilman, 1990).

The relative importance of interval- versus event-dependent dynamics has been tested by both experimental and comparative studies. In most fire-prone vegetation, event-dependent effects are at least as important, and often more important, than interval-dependent effects. Interval-dependent effects are deterministic and predictable. Event-dependent effects are generally far less predictable and hostage to the vagaries of weather and fire behaviour. Event-dependent effects cause intrinsic uncertainty in fire-driven dynamics. The levels of uncertainty are greatest in woody vegetation with heterogeneous fuels, and least in grasslands with more homogeneous fuels.

The theme of event- versus interval-dependent regulation will recur throughout the book. Interval-dependent effects are those most familiar from general ecological theory. They concern growth rates, mortality and fecundity and their influence on species performance in competitive encounters. In many fire-prone systems, population changes depend less on competition and more on the interplay between fire and attributes governing persistence and recruitment.

1.6.4 The evolutionary ecology of fire

Evolutionary questions have been surprisingly neglected by fire ecologists. Are there any plant traits peculiarly associated with fire? How does fire shape life history, morphology and reproduction? Has fire been significant in the evolution of whole floras? Life-history evolution of fire-prone populations is guided by the principle of allocation – that plants allocate resources in such a way that fitness is maximized. Recent studies have shown that sprouting carries a considerable allocation cost which results in reduced seedling growth rates, deferred reproduction and smaller seed crops (section 6.2). The evolution of non-sprouters from sprouting ancestors has been explained in terms of these allocation costs. By foregoing allocation of carbon to reserves for sprouting, non-sprouters gain early growth and reproductive advantages relative to juveniles of sprouters. However, these traits are advantageous only in large gaps where there is sufficient space. The distribution of the two alternative life histories across moisture gradients and in areas experiencing different fire frequencies has been explained in terms of the availability of recruitment gaps (section 6.2.3).

Allocation costs may also underlie selection for age to maturation in both sprouters and non-sprouters (Clark, 1991). Clark has modelled optimal solutions for reproductive scheduling in relation to the increasing risk of fire with stand age (section 6.3). He assumed that the optimal life history for non-sprouters was that which maximized the probability that a plant will be reproductively mature at the time of the next fire. For sprouters, the optimal life history was that which maximized the number of recruitment episodes experienced over the reproductive life of the plant. This differs from conventional life-history theory which explores ways of maximizing net reproductive rate in continuously growing populations (Stearns, 1977; Caswell, 1982). Clark showed that life histories of two pine species were broadly consistent with predictions based on these two principles. This emphasis on persistence of populations rather than their density is implicit in many studies of fire-dependent plant population dynamics (Chapter 5).

The evolutionary ecology of fire adaptations is best developed for reproductive traits. Serotiny, the store of seeds in insulated structures in the canopy, has received the most attention. Serotiny has been shown to be genetically determined and is a polymorphic trait in some North American conifer populations. The frequency of serotiny in a population has been directly related to the time since the last large fire (section 6.4.1). Serotiny and other fire-stimulated traits have some analogies to mast-fruiting trees in forests and have also been interpreted as devices to satiate predators.

One of the most controversial evolutionary questions is whether plants have evolved traits that enhance their flammability (section 6.6). Mutch (1970) argued that some communities were more fire-prone than others because they had evolved traits which led to an increased risk of burning. Enhanced flammability was seen as a device for eliminating more competitive species that would replace fire-prone communities in the long absence of fire. Mutch's observations are consistent with many fire-prone communities world-wide. Without fire, they are replaced by less flammable communities which seldom burn. Part of the difficulty with the theory has been that is was couched in group-selectionist terms. A recent model has explored conditions under which enhanced flammability might evolve for individualistic benefits (Bond and Midgley, 1995). Results of the model suggested that flammability can evolve if it causes less flammable neighbours to burn, creating gaps. However, to outcompete less flammable genotypes in these gaps, flammability would have to confer additional benefits such as increased growth rates. Many of the traits that influence flammability are likely to confer such additional benefits. However, it might prove difficult to separate the relative importance of fire and other factors in their evolution.

The evolution of flammability in just a single common species could trigger cascading effects on ecosystem composition by causing an increased occurrence of fires. Evidence for such effects on the evolution of floras has seldom been explored (section 9.2.4). The occurrence of fire-recruiting species in Mediterranean shrublands suggests that these shrublands may owe part of their character to fire, and not to convergent climates. The evolution of grasses, many with highly flammable fuel properties, must have had devastating effects on tree floras, with only a handful of species able to tolerate the new conditions imposed by fire. Distinct savanna floras with endemic or near endemic woody genera occur in African and South American savannas but not in the savannas of Madagascar which are thought to be largely of anthropogenic origin. There is much scope for more evolutionary analyses, both of the evolution of species traits and the evolution of floras.

1.6.5 Fire and plant communities

(a) Succession

Studies of the ecology of fire-prone communities has taken a different course from the mainstream of plant community ecology, similar to the different emphases seen in population studies (section 7.2). This is most clearly seen in studies of plant succession. In communities little affected by fire, elegant theory, tested by experiment, has been developed for predicting species replacement sequences on the basis of their relative competitive ability. Species changes over time can be predicted on the basis of their position in the competitive hierarchy. The hierarchy, in turn, can be explained in terms of different patterns of resource uptake and allocation (Tilman, 1982, 1988). A different kind of succession has been emphasized in fire-prone communities (section 7.2). Change is driven by the life-history attributes of species, not competitive ability. Species turnover reflects intrinsic differences in growth rates and longevities, not their position within a competitive hierarchy.

This model is non-interactive. It assumes that species would behave in the same way regardless of the species mix. Even if a short-lived, early successional species is suppressed by a late successional (long-lived) species, it can persist in the community in the form of a dormant seedbank until the next burn. Multiple successional pathways can occur as a result of the interplay between the fire regime and the life-history attributes of species occupying a site. Traits considered important for predicting successional change are quite different from those used in competitively driven succession (sections 5.10 and 7.2.5). They relate to individual persistence and recruitment requirements of a species in relation to the frequency and severity of fire. Species interactions in the interval between fires are of peripheral interest. This view of succession has arisen from many studies of succession in fire-prone shrublands and woodlands. Qualitative models have been developed for predicting successional pathways and are fair reflections of patterns observed in nature. However, the assumption that species interactions between fires are unimportant has seldom been tested.

Neither life-history attributes nor competitive hierarchies seem particularly useful in interpreting the dynamics of grass/tree mixtures in savannas. Savanna dynamics involve complex interactions between fluctuating climate, herbivore pressure, rainfall, human impacts and patchy soils. Competitive hierarchies change with plant size. Juvenile trees are prevented from emerging to the canopy by interactions with grasses, including their effect on fire. However, once they have escaped to reach the canopy, trees can suppress the grass understorey and reduce the frequency of burning. In the long absence of fire, savannas can be converted to non-flammable forests or thickets. The existence of this intrinsically unstable growth-form mix over 20% of the Earth's surface has not been adequately explained.

(b) Fire and spatial patterns

Although fire succession has been well studied, much less attention has been paid to the impact of fire on community composition. Small-scale heterogeneity within communities is an important component of community diversity. There is growing evidence that heterogeneity in burning patterns creates fine-scale heterogeneity in communities (section 7.3.1). Communities with patchy fuels create patchy fire intensity which causes patchy fire survival and heterogeneity in post-burn recruitment. We are only just beginning to explore the impact of these kinds of processes on coexistence and diversity patterns in fire-prone communities. They may prove analogous to patchy nutrient distribution in creating opportunities for coexistence in communities (Tilman, 1982; Auerbach and Shmida, 1987).

At larger, landscape scales, fire movement and fire properties are strongly affected by topography and natural firebreaks. Community turnover along resource gradients can be attributed partly to fire movement patterns. This is most obvious in the restriction of non-flammable communities (such as forests) to landscape facets that provide refugia from fire ((Plates 5 and 6) section 7.3.2). Consistent differences in the fire regime, due for example to topographic location, may also explain community turnover within fire-prone communities (Plate 4). This scale of fire-patterning has been poorly explored. However, there are several examples where the distribution limits of species have been shown to be determined by the fire regime, rather than by physiological constraints or competitive interactions (section 7.3.2).

(c) The composition of communities

Studies of competition have been neglected in fire-prone communities. However, competition may be important in determining the number and composition of species occupying similar fire-successional niches. Recent studies of dominant shrubs in South African, Australian and Californian shrublands have begun to explore how competition may account for variation in community membership. These suggest that competition is not unimportant in determining the mix of co-occurring species, especially those with similar growth forms. Some studies suggest that competitive sorting may begin very early after a burn.

(d) Species diversity

Competition is also important, along with disturbance, in determining species diversity. The intermediate disturbance hypothesis predicts peak diversity at intermediate disturbance levels. Long intervals without disturbance will result in sufficient time for the exclusion of species by strong competitors. Very short intervals could results in elimination of all species but those most resilient to disturbance. A number of studies have reported a monotonic decline in the number of species with increasing stand age after a burn. However, this is not a test of the intermediate disturbance hypothesis. Species may appear to be lost as a stand ages but persist in dormant seedbanks to reappear after the next fire. Diversity patterns within a fire cycle should not be confounded with changes in diversity caused by variation in the length of each cycle. Very few studies seem to have recognized the problem and no clear picture has emerged on diversity/fire frequency relationships (section 7.5).

1.6.6 Fire and management

In Chapter 8, we address briefly the use of fire in managing communities. There are many books on the topic for different regions of the world. In this book, we only discuss guiding principles, and key constraints, determining fire management. Models of resource management have generally been based on deterministic successional pathways and have assumed predictable replacement sequences. These are analogous to the interval-dependent models of populations and communities, which emphasize competition as the engine driving rangeland dynamics. New models, developed in fire-prone communities, take much greater account of event-dependent effects (Westoby *et al.*, 1989). Vegetation can switch abruptly to a new state depending on the set of external conditions. Recently, a framework for mathematical modelling of these 'state-transition' models has been developed which offers much promise for exploring management options (Starfield *et al.*, 1993).

1.6.7 Fire and scale in ecology

Considerable progress has been made in understanding the effects of fire at a population level. How well do these reproduce landscape and global scales of fire effects? Fire regimes have undergone enormous changes due to the growing impact of humans. Fire suppression, sometimes on a massive scale, together with the addition of non-native species, has changed fuel conditions. The continuity of fires has been fragmented in some landscapes and increased in others. Humans have become the major ignition source in many vegetation types. Global warming, triggered by the use of fossil fuels, is changing the climate setting and thus the kind of fire regime. To understand the impact of these factors one needs to combine reductionist studies at the population level with much larger spatial scales.

Contemporary correlations between climate and vegetation have been used effectively to indicate the scale of vegetation changes that we can expect with global warming (section 9.2.1). However, correlative models assume direct climate control of plant distribution, and evidence for this has proved elusive at population level (Woodward, 1987). Climate may influence the distribution of vegetation through its indirect effects on fire. Palaeoecological and contemporary studies have shown that fire regimes closely track climate (section 9.2.3). Permanent changes in plant cover can happen abruptly due to the effects of a single severe fire, as several studies of boreal vegetation have shown. These studies suggest that extreme fires may be the immediate cause of vegetation change in some communities in a warmer world.

Gap models are currently a popular tool for relating population-level phenomena to global changes in vegetation pattern (section 9.4). These models trace the growth of individual trees in a stand (they have also been extended to grassy formations). Methods have been devised for extrapolating the results to wider spatial scales. The models are appealing because they can incorporate physiological information of increased CO_2 and increased temperatures through the effects on individual tree performance. Gap models emphasize interval-dependent effects. The relative growth of individuals determines tree mortality and species replacement patterns. Gap models can also be used to model the effects of disturbance by incorporating vital attributes that are important for persistence under a given disturbance regime. However, gap models may be unnecessarily complicated for predicting the effects of changing fire regimes. In some communities, simpler approaches, based on the probability of persisting under changing fire regimes, may be more appropriate. The problem of scaling up from population to landscape under changing fire regimes is a major challenge for future research in fire ecology.

2 Why and how do ecosystems burn?

There is an enormous literature on the loss of plant parts to insect and vertebrate herbivores and the morphological and chemical plant defences that have developed to counteract these losses. Herbivores are thought to exert a major impact on plants, both in ecological and evolutionary time scales. Annual losses to herbivores are estimated at more than 10% of plant production in natural communities, which probably exceeds average allocation to plant reproduction (Coley *et al.*, 1985). Fires, by comparison, can remove 80% of the above-ground primary production in many ecosystems. Examples include some grasslands (where most of the grass biomass can be burnt annually) or fynbos shrublands (with above-ground primary productivities around $250\,g\,m^{-2}\,yr^{-1}$, and fires burning around $3000\,g\,m^{-2}$ once every 15 years).

Fire is an extremely effective 'herbivore' in many plant communities in the world. Fires consume 'fuel', a broader diet than biotic consumers, incorporating flammable leaves, shoots and litter. Just as the chemical and morphological properties of leaves and shoots make them more or less susceptible to herbivory, so too some kinds of plants are more susceptible to fire than others. Unlike herbivory, however, the 'bite' of fire falls much more evenly across species within a community. In addition, the extent of plant losses to fire differs greatly between communities. Ironically, plant characteristics considered to be adaptations to reduce animal herbivory (such as tough, fibrous, long-lived leaves, Coley *et al.*, 1985) promote 'herbivory' by fire, while the most palatable, weakly defended leaves of highly productive environments reduce the likelihood of consumption by fire. In this chapter, we consider the environmental conditions that promote fire and the plant attributes that make vegetation flammable, and attempt to place the many studies of plants as 'fuel' into context for community and evolutionary ecologists.

2.1 THE FIRE REGIME

Fire, rather like insect outbreaks, has an irregular incidence. This irregularity is important in accounting for differences in the effects of fire. Fire varies in terms of how often it occurs (**frequency**), when it occurs (**season**) and how fiercely it burns (**intensity**). Combinations of these three elements are termed the fire regime. Fires also vary in their **type**: **ground** fires burn underground, in organic layers of the soil; **surface** fires burn just above the ground surface, for example in grass swards in savannas or in litter layers below coniferous trees (Plate 7); and **crown** fires burn in the canopies of trees (Plate 8). Usually the latter are sustained by the surface fires below them (dependent crown fires), but under severe conditions crown fires can race ahead of the surface fires (independent crown fires). Yet another source of variability is whether fires burn with the wind or upslope (headfires) or against the wind or downslope (backfires).

Both the mean and the variability around the mean for each element of the fire regime is very important. Plant species tend to be specialized to a particular frequency, season or intensity of fire. Changes in fire regime, or spatial variations in intensity (Hobbs and Atkins, 1988) can favour one species over another (Chapters 4 and 5). Variability in the fire regime may influence species coexistence should successive fires be sufficiently different to favour different species on the same site (Chapter 7).

2.2 PREREQUISITES FOR FIRE

In order for fires to occur, three conditions must be met. The environmental (weather) conditions must be such that the fuel is dry enough to burn, some source of ignition must be present, and there must be sufficient fuel. These conditions vary considerably between ecosystems, resulting in distinct fire regimes (Table 2.1). The incidence of fires and their behaviour in vegetation can be predicted by models of the interactions between these three conditions. Such models have been developed with the purpose of managing 'the fire problem' in many areas, and for predicting the behaviour of fires, but they offer community ecologists some insights into how and why ecosystems burn.

2.2.1 Weather and climate

The climate of a region is usually described in annual or seasonal terms; weather, on the other hand, refers to day-to-day, or even hour-to-hour, changes in atmospheric conditions. Weather is an important determinant of how and when fires will burn. Its effects can be both direct and indirect. Wind affects the spread and intensity of fires directly. Rainfall, relative humidity and temperature all affect fire behaviour indirectly through their effect on the moisture content of vegetation and litter. These effects are relatively easily described and modelled. The dynamics of the atmosphere are less easily modelled, but are very important. For example, the stability of the atmosphere can modify fire behaviour, and determine whether a surface fire will become a crown fire, or whether spotting (the establishment of new fires ahead of the main fire by firebrands carried aloft) will take

Table 2.1 Examples of fire regimes for important vegetation types world-wide; the examples are drawn from non-modified systems

Vegetation type	*Frequency of fire*	*Season of fires*	*Intensity of fires*
Grasslands	Annual or longer, depending on grazing pressure and rainfall which determine fuel load	Fires occur in the dry season when grasses are dormant	Fire intensities range from <100 to >5000 $kW\,m^{-1}$
Californian chaparral shrublands	Fires occur at intervals of between 25 and 100 years	Fires concentrated in dry summer periods	Fire intensities can be high (>50 000 $kW\,m^{-1}$)
South African fynbos shrublands	Fires occur at intervals of between 5 and 40 years	Fires concentrated in dry summer periods	Fire intensities range from 500 to 30 000 $kW\,m^{-1}$
African savannas	Frequencies range from annual to once every 30 years or more, depending on rainfall and grazing pressure	Fires occur in dry seasons when grasses are dormant	As for grasslands
Brazilian cerrado	Frequencies of 1–3 years	Fires occur in dry winters	No data?
Australian eucalyptus woodlands	Some areas burn frequently, even annually (surface fires); stand-replacement crown fires can occur every 100–300 years in wet sclerophyll forests	Dry season	Usually low or moderate intensity (500–3000 $kW\,m^{-1}$); high-intensity crown fires (7000–70 000 $kW\,m^{-1}$) are rare
North American coniferous forests	Surface fires at frequent intervals (1–10 years). Crown fires at intervals of 100–1000 years	Dry summer periods	Low intensity (200–800 $kW\,m^{-1}$) for surface fires, but very high (>50 000 $kW\,m^{-1}$) for crown fires (stand replacement fires)
South American rainforests	Fires very infrequent and restricted to single trees or small patches	Only possible after several rainless days	Very low intensities (<20 $kW\,m^{-1}$)

place or not. Pyne (1984) gives a detailed account of fire weather, which will not be presented here.

Fires can occur under most climate regimes, provided dry conditions occur at some time of the year. Fire does not occur in Antarctica (which is too wet and cold, and has no fuel), and rarely (but not never) in deserts (not enough fuel) or rainforests (too wet). Climate also affects the occurrence of fires through its effects on primary productivity, especially in drier areas with variable rainfall. Fires in African savannas, for example, may only burn in years of above-average rainfall when sufficient grass biomass accumulates and outstrips the ability of grazers to prevent the build-up of fuel.

2.2.2 Sources of ignition

Fires will only occur if some source of ignition is present. Lightning is the chief cause of fires in natural vegetation. In fynbos shrublands of the Cape mountains in South Africa, for example, about half of all recorded fires were initiated by lightning (Horne, 1981). Sparks from hardened quartzite rocks are also often cited as a cause of fires in the mountains of the Cape (Kruger and Bigalke, 1984). Fires caused by active volcanoes occur in some parts of the world. However, humans cause the majority of contemporary fires. There is direct evidence that hominids 'domesticated' fire over 1 million years ago (Brain and Sillen, 1988). The spread of colonial powers from the relatively fire-free landscapes of Western Europe to the western and southern hemisphere took an abhorrence of fire with them (Pyne, 1990) which led to active fire suppression in many parts of the world. Since the 1960s, however, there has been a growing acceptance of the role of fire in ecosystems, and of prescribed burning (Biswell, 1989). With increasing population levels, the relative importance of humans as a source of ignition will continue to increase. Modern fire regimes that are unaffected by human sources of ignition are today extremely rare.

2.2.3 Fuel

Finally, the occurrence of fire in natural ecosystems depends on enough plant material being present to support a fire. This is not simply a question of primary productivity. The complex relationships between different components of vegetation structure determine whether ecosystems of similar productivity will be fire prone or fire resistant. These relationships are considered in some detail in the next section.

2.3 PLANTS AS 'FUEL': WHAT MAKES VEGETATION FLAMMABLE?

The fuel properties of vegetation, that is the particular mix of characteristics that determine whether, and how, it will burn, can vary considerably between ecosystems. Both the morphological and chemical properties of individual species and the arrangement of many species and communities in space influence their susceptibility to fire. Morphological and chemical features are often associated with quantitative defences against herbivory and adaptations to surviving drought, but they seem to promote the chances of fire. The study of plant structure in relation to fire originates from the needs of foresters and range managers to be able to predict the behaviour of fires (Dell and Philpot, 1965; Countryman and Philpot, 1970; Philpot, 1977; Rundel, 1981). However, the results of this work can indicate how the structural properties of plants make them more flammable.

2.3.1 The structure and chemistry of individual plants in relation to fire

Individual plants possess properties that make them more or less susceptible to fire. The most important of these is the production or retention of dead material. The moisture content of dead plant parts is usually low (between 5 and 15% in dry weather, expressed as a percentage of dry mass), and it is dead fuel rather than live that initially carries fire. A fire will not normally be sustained if there is not enough dead material, or should the dead material be too moist. Combustion of dead fuels drives the moisture out of living fuels. Living plant parts initially act as a heat sink, but once they have dried to a point where they begin to burn, they contribute to energy release in fires. Live plant parts have moisture contents between 50 and 250% or greater (expressed as a percentage of dry mass). Live leaves will burn more easily if their moisture content is low, as less water needs to be driven out of them prior to ignition. Thus leaves well defended against herbivores, with high fibre contents and high leaf specific weights (Coley, 1988; Reich *et al.*, 1991), will tend to burn more easily because of their lower moisture content.

Besides the amount of dead and live material produced, the shape, size and arrangement of plant parts has a major influence in determining susceptibility to fire. Surface area to volume ratio is important both because it influences the rate of moisture loss (small-diameter plant parts lose moisture more rapidly than larger ones), and because it affects the ratio of fuel to air (the packing ratio, section 2.3.2). Examples of fire-prone species include pines, grasses and heaths, with their finely divided leaves. On the other hand, rainforest trees and cacti have large, bulky leaves (or spiny leaves with bulky stems) which render them less prone to fire (Figure 2.1).

Fire behaviour modellers have recognized the importance of particle size, when describing the fuel properties of plants, by separating fuel into size classes. These were based on rates of water loss from dead organic materials, which declines logarithmically with time (Pyne, 1984). Size classes (or 'timelag classes') are based on the time for a dead fuel particle to

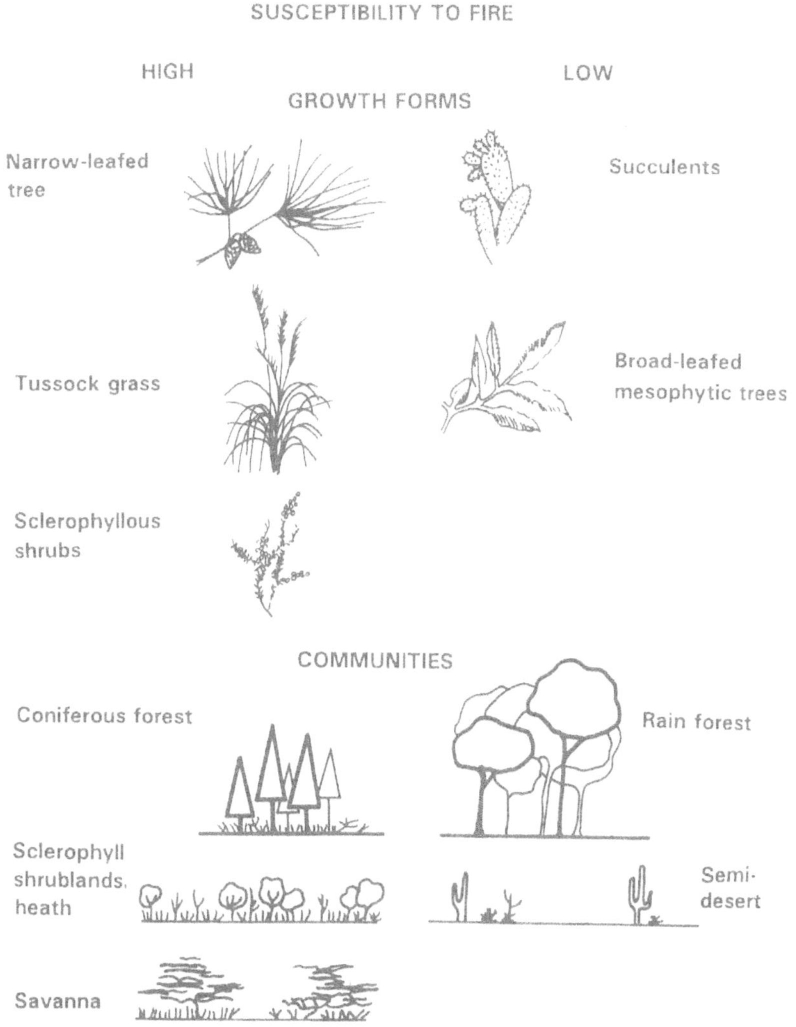

Figure 2.1 Flammable plants are characterized by relatively low moisture contents and fine plant parts arranged so as to form a fuel–air matrix that optimizes propagation of fire. Less-flammable species have relatively high moisture contents and relatively coarse dimensions. Fire-susceptible communities are made up of flammable species and accumulate well-aerated fuel beds with relatively large quantities of fine dead material. Fire-resistant communities have a lack of such fuel beds (e.g. forests with densely packed litter layers that decompose quickly or arid ecosystems with discontinuous fuel beds).

lose 63% of the difference between its initial moisture content and a new equilibrium moisture content under standard laboratory conditions of 27°C and 20% relative humidity (Byram, 1963). The timelag classes used to partition dead fuels are: 1h (0–6mm), 10h (6–25mm) and 100h

(25–75 mm) (Deeming and Brown, 1975). Environmental selective pressures favouring small or narrow, cylindrical leaves and thin twigs will increase the amount of 1 h fuels and promote flammability.

A further factor that may influence the combustion properties of plant material is the energy content, or heat of combustion. The heat of combustion refers to the total amount of energy contained per unit mass of fuel. Not all of this energy is released during a fire. The actual amount of heat released during a fire decreases with increasing moisture content of the fuel, and is termed the heat yield. Byram (1959) gives correction factors for determining heat yield from heat of combustion and moisture content values (Figure 2.2). Heat yields are important in calculating fire intensity. The most commonly used measure of fire intensity is Byram's (1959) fireline intensity:

$$I = Hwr$$

where I is the fireline intensity ($kW m^{-1}$), H is the heat yield of the fuel ($J g^{-1}$), w is the mass of fuel consumed ($g m^{-2}$), and r is the rate of spread of the fire ($m s^{-1}$). Heat of combustion values do not vary a great deal between plant species, and are probably much less important in determining flammability than mass, structure and moisture content. For example, heat

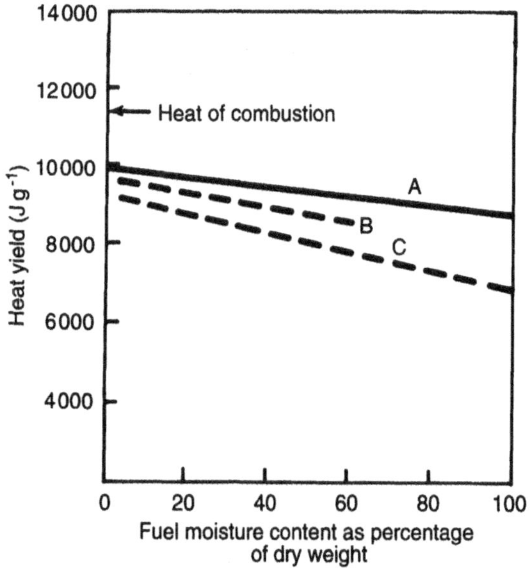

Figure 2.2 The relationship between heat of combustion, moisture content and heat yield in wildland fuel complexes (after Byram, 1959). Heat of combustion values (usually determined in a bomb calorimeter) have to be reduced for moisture content to calculate the heat yield (the energy released during fires). Lines A, B and C show estimates of heat yield for complete combustion, small fires and large fires, respectively.

of combustion values for plants are usually somewhere between 18000 and 21000 J g^{-1}; rates of spread may range from less than 1 to over 100 m min^{-1}, while the amount of fuel consumed ranges from 200 g m^{-2} in grasslands to 20000 g m^{-2} in some coniferous crown fires. The influence of moisture content on the heat yield introduces further variability to the equation (Figure 2.2); given these differences, heat of combustion almost becomes a constant when calculating fireline intensity. However, plants that do have high levels of oils, fats, waxes and terpenes (or simply crude fats) may be significantly more susceptible to fires. Crude fats are volatile, and make plants more flammable, because they are driven out of live plants at relatively low temperatures, and then burn as gases. In addition, oils and waxes have up to twice the heat of combustion values of other plant material (Philpot, 1969), thereby increasing the intensity at which they burn. Many of these secondary compounds are viewed as antiherbivore defences, but while, on the one hand, they deter herbivores, they encourage fires on the other.

2.3.2 The properties of vegetation communities in relation to fire

In simple communities dominated by relatively few, or even a single species, such as some conifer woodlands and chaparral shrublands, the properties of an individual plant species (as described above) will determine the likelihood of fire consuming the whole community. In species-rich communities, however, the combined properties of all species on a site determine whether and how the vegetation will burn. The fuel properties of communities are less easily described than those of species. Most studies of plants as fuel have concentrated on single species, or have been restricted to one or a few community descriptors, such as biomass. Early fire behaviour prediction studies were empirical, and used either community biomass (McArthur, 1966, 1967) or the spatial arrangement of fuels as inputs (van Wagner, 1977). Models of fuel accumulation and decay (Walker, 1981) were developed to estimate community litter mass to support the McArthur models (Kessel *et al.*, 1982). The development of physical fire-spread models (Rothermel, 1972; Kessel *et al.*, 1978) led to the need to describe communities more fully (see below), but still the data were drawn from sometimes extensive studies of the fuel properties of single species. A good example is the chaparral shrub *Adenostoma fasciculatum* (chamise). This shrub is often the only dominant species in many Californian chaparral stands, and has been the subject of intensive study (Dell and Philpot, 1965; Countryman and Philpot, 1970; Rothermel and Philpot, 1973). However, the various approaches used in fire-spread models are now being used by ecologists as a basis for understanding how and when communities will burn (Kauffman *et al.*, 1988; van Wilgen *et al.*, 1990a). They could also be useful for predicting the effects of fire based on a knowledge of the response of species to fires of different intensity.

When considering the fuel properties of vegetation communities, a good start would be to examine the dynamics of dead material. The occurrence of fire in a plant community depends on dead material, which accumulates according to the rate of decay. The balance between accumulation and decay can be expressed as:

$$X_t = L/K\,(1 - e^{-Kt}) \quad \text{where} \quad L = KX_{max}$$

where X_t is the amount of dead litter fuel at time t (in years), L is the litter production rate (tonnes ha^{-1} yr^{-1}), X_{max} is the maximum litter standing crop at equilibrium, K is the decomposition constant, and e is the base of the natural logarithms. The model assumes steady rates of input and decomposition. Given that post-fire age, and litterfall and decomposition rates are known, it would be possible to calculate litter fuel loads. Equilibrium fuel loads (X_{max}), which occur when litterfall is balanced by decomposition, vary considerably for different ecosystems (Table 2.2, Figure 2.3). Foresters have used this approach to predict fuel loads for the management of fire suppression (Kessel *et al.*, 1984), but it also tells ecologists that communities with high decomposition rates will have lower fuel loads and will be less likely to burn. In rainforests, for example, decomposition rates are high, resulting in relatively low steady-state litter loads. In South African fynbos, on the other hand, decomposition is very slow, and large amounts of litter accumulate, increasing the likelihood of fire.

Table 2.2 Examples of litterfall, decomposition and steady-state litter fuel loads for selected vegetation communities

Vegetation	*Region*	*Litterfall* ($gm^{-2}yr^{-1}$)	K	*Steady-state litter load* (gm^{-2})	*Source*
Grassy woodlands	Northern Territory, Australia	250	0.42	600	Walker (1981)
Heathlands	South Australia	90	0.09	950	Walker (1981)
Tropical rainforest	Queensland, Australia	511	0.91	562	Walker (1981)
Chaparral shrublands	California	355	0.23	1543	Schlesinger and Hasey (1981); Mooney *et al.* (1977)
Fynbos shrublands	Cape, South Africa	350	0.07	5000	Mitchell and Coley (1987)

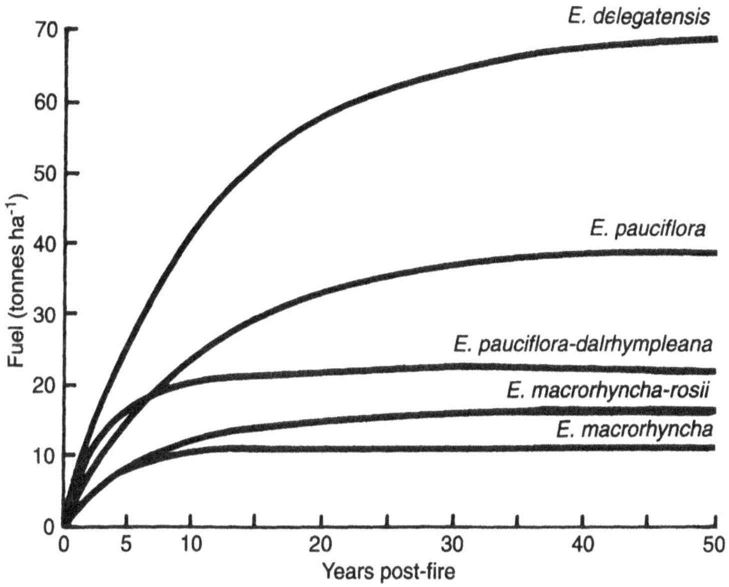

Figure 2.3 Variations in accumulated fuel load over time with differing decomposition rates in some Australian *Eucalyptus* forest and woodland communities (after Kessel *et al.*, 1982).

Besides the amount of fuel, it is necessary to consider the spatial structure of whole communities. The arrangement of plant parts in space can be summarized, for the purposes of predicting fire behaviour, as packing ratios. These are defined as the ratio of bulk density (fuel mass per unit volume occupied by fuel and air) to the specific gravity (fuel mass per unit volume occupied by fuel alone) (Rothermel, 1972; Pyne, 1984). This is essentially a measure of the fuel:air mix. As is the case with internal combustion engines, there is an optimum fuel:air mix that will maximize combustion. This can be extremely important in determining whether vegetation communities will burn or not.

The complex relationships between biomass, spatial structure and fire behaviour drove foresters and engineers to develop the concept of fuel models to describe communities. The best-developed examples of this approach are the fuel models used as inputs to Rothermel's (1972) fire model. The fire model is designed to predict fire behaviour in a single continuous stratum of fuel contiguous to the ground. The fuel models (Deeming and Brown, 1975) summarize the physical and chemical make-up of the community, and include:

- fuel loads divided into live herbs, live woody fuel (<6 mm), and dead (1, 10 and 100 h) fuel (section 2.3.1);
- surface area to volume ratios for each of the above categories;
- the fuel bed depth which, combined with surface area to volume ratios, provides an estimate of packing ratio; and
- a mean fuel energy content for the combined fuel mass.

Standard examples of these models for several vegetation types in the United States have been developed (Table 2.3), as well as software to assist users in the preparation of custom-built fuel models (Burgan and Rothermel, 1984). Fire behaviour predictions from these models illustrate the role of fuel structure in determining fire behaviour (Table 2.3). Under identical weather conditions, chaparral burns with longer flames and higher intensity than grasslands, due mainly to much higher fuel loads. Coniferous forest understoreys, on the other hand, burn with smaller flames and at lower intensity than grasslands, despite much higher fuel loads, because of the dense packing ratio of the fuel bed, reflected in the low fuel bed depth. This illustrates the importance of fuel arrangement, and that fuel load alone is insufficient to determine fire behaviour.

Table 2.3 Examples of fuel models developed for fire behaviour prediction in three vegetation types in the United States, with estimates of fire behaviour under identical conditions[a]

	Short grass	*Chaparral*	*Coniferous forest understorey*
Fuel loads (tonnes ha^{-1})			
Dead fuel			
1 h	1.7	11.2	6.7
10 h	0	9.0	4.5
100 h	0	4.5	11.2
Live fuel			
Herbaceous	0	0	0
Woody	0	11.2	4.5
Surface area to volume ratios (cm^{-1})			
Dead fuel			
1 h	115	66	66
10 h	–	35	35
100 h	–	9	9
Live fuel			
Herbaceous	–	–	–
Woody	–	49	49
Heat of combustion (J g^{-1})	18595	18595	18595
Fuel bed depth (m)	0.3	1.85	0.3
Fire behaviour			
Flame length (m)	1.5	6.5	0.3
Intensity (kW m^{-1})	666	15293	162

[a] Conditions for estimating fire behaviour were: dead fuel moisture 8%; live fuel moisture 100%; wind 10 km h^{-1}.

2.3.3 What factors predispose communities to fire?

Communities will burn if they contain one or many of the 'right' properties. These properties include enough fuel, low moisture contents in the live leaves, an optimum arrangement of plant parts in space, low decomposition rates and high levels of secondary compounds. The accumulation of sufficient biomass to support a fire requires relatively high levels of primary productivity (deserts do not burn), and low levels of herbivory and decomposition to ensure litter accumulation. Thus one of the reasons that rainforests do not burn is that they experience a high turnover of litter, and do not accumulate much dead fuel (Table 2.2). High levels of herbivory in fertile savannas and grasslands effectively reduce the occurrence of fire in many areas. In contrast, infertile savannas and grasslands are not subject to extensive herbivory, accumulate fuel and burn. Mediterranean shrublands also experience extremely slow decomposition and low levels of herbivory when compared to other vegetation types, despite high rainfall, and fire inevitably results (van Wilgen and van Hensbergen, 1992).

Low moisture contents (expressed as a percentage of dry mass) in live leaves in vegetation mean that communities will be more susceptible to fire. Californian chaparral shrubs, for example, show strong seasonal trends in moisture content, with low (50%) moisture contents at the height of the fire season (Dell and Philpot, 1965). This makes it possible for them to burn despite the lack of an herbaceous understorey. Fires also burn in the crowns of Australian *Eucalyptus* woodlands, and in many North American conifer forests, where foliar moisture contents range between 70 and 130% in the dominant trees. Relatively high moisture content (140–200%) in the leaves of broad-leaved Canadian aspen, birch and maple trees, on the other hand, effectively prevent the onset of crown fires (van Wagner, 1977).

Although biomass and moisture content are important, the spatial arrangement of plants appears to be of overriding importance in determining whether vegetation will burn or not. Examples of fire-prone communities (Figure 2.1) include coniferous forests, which burn because the crowns of trees are densely packed and have low moisture contents, and because litter layers accumulate due to low levels of decomposition; fynbos shrublands with high levels of standing dead material, and many species with fine leaves; and savannas with erect grass swards below the trees. Fire-resistant communities include tropical rainforests (Figure 2.1) which do not burn because of their spatial structure, despite high fuel loads. Rainforests are dominated by large trees with relatively sparse canopies of live foliage. Dead material is restricted to the transient litter layer below the tree canopies, preventing the occurrence of fires (van Wilgen *et al.*, 1990a). Van Wagner (1977) proposed a model for predicting conditions that would lead to crown fires in coniferous timber stands. The model assumes that crown fires are dependent on surface fires for active spreading. In order for crown fires to develop, surface fires have to reach a critical intensity (I, $kW\,m^{-1}$)

and rate of spread (R, $m\,min^{-1}$), which varies according to crown bulk density (CBD, $kg\,m^{-3}$), foliar moisture content (FMC, %) in the crowns of trees, and the lower crown branch height above the ground (LCBH in metres). The critical intensity necessary is given by:

$$I = [(0.01\ LCBH\ (460 + 26\ FMC)]^{1.5}.$$

The critical rate of spread is given by:

$$R = 3/CBD.$$

A sensitivity analysis reveals that forests with low crown branch height, low fuel moisture content and high crown bulk density would require less intense surface fires to initiate the onset of a crown fire. These attributes would therefore indicate susceptibility of a community to fire. In addition, 'bridging fuels', which carry fires from the forest floor to the canopy, would promote the onset of crown fires (van Wagner, 1977).

Finally, vegetation communities that have high levels of secondary compounds (for example *Eucalyptus* woodlands) may be more prone to fire, or may burn with greater intensity than similar vegetation without such compounds (Mutch, 1970).

2.3.4 Fire regimes in different communities

The fuel properties of vegetation communities combine with environmental conditions to produce different types of fires (Table 2.4). Only surface fires are possible in grasslands and savannas, and these vary in intensity depending on the conditions under which they burn. In savannas, trees react to the intensity of fires in the grass layer, and the structure and composition of the tree layer can be largely determined by fire intensity. Fires in heathlands and shrublands are essentially crown fires in the canopies of the shrubs, and can be of much higher intensity than grass fires due to higher fuel loads in many cases. These fires are almost always 'stand-replacement' fires, where adult shrubs are killed or burnt back to ground level. In coniferous forests, the situation is more complex. Surface fires often burn in the litter layers below the trees under mild conditions, and these low-intensity fires may have little or no effect on the overstorey trees (Plate 7). Under drier conditions, the fires can burn into the crowns of the overstorey trees, but they will be dependent on the surface fires below them; many trees will be killed and the resultant gaps in the canopy will allow for regeneration to take place. True 'stand-replacement' fires will occur under severe conditions, when running crown fires of exceptionally high intensity can set the successional clock back to zero (Plate 8). Finally, tropical rainforests rarely burn, due partly to the lack of suitable dry weather conditions, but also to the fuel properties of these areas which hinder the development of fires.

Table 2.4 Fuel characteristics and resultant types of fire in major vegetation groups of the world

Vegetation type	*Fuel characteristics*	*Types of fire*
Semi-desert	Sparse plants	Fires rare or absent due to sparse fuel
Grassland	Fine grass litter and live grass	Surface fires
Savanna	Fine grass litter and live grass. Trees not normally considered part of the fuel complex	Surface fires. Tree layer affected by intensity of surface fire
Heathlands and shrublands	Fine-leaved shrub fuels with varying amounts of dead material. In some cases (e.g. fynbos) herbaceous fuels co-occur	Fires in crowns of shrubs
Tundra	Sparse plants	Fires rare or absent due to sparse fuel
Coniferous forest	Stratified fuel beds consisting of needle litter, twigs, cones and understorey shrubs below a canopy of live foliage	Surface fires under mild conditions. Under hot, dry conditions, dependent crown fires develop. Under severe conditions, running crown fires will occur. Huge differences in fire intensity between these types
Tropical rainforests	Stratified fuel beds with leaf litter layer, understorey shrubs and trees, and canopy trees	Fires rare or absent due to moist conditions, transient litter layers and high moisture content of leaves in canopy trees

2.4 FIRE RECURRENCE INTERVALS AND THEIR MEASUREMENT

One of the most important questions asked by community fire ecologists relates to the frequency of fire. How is it possible to describe this feature of the fire regime, given that fire return intervals may be measured in decades or even centuries? Early work of this nature was confined to forested ecosystems, and relied heavily on an analysis of fire scars on trees (Spurr, 1954). However, the need for a deeper understanding of disturbance and its implications for community dynamics and land management has led to the development of quantitative methodologies (Johnson and van Wagner, 1984). Heinselman (1973) laid a basis for this methodological development by mapping past fires as carefully as possible, using written records, fire scars, and stand origins. This work, and other studies conducted elsewhere, has led to the development of stochastic models describing the fire history of sites (Johnson, 1978; Johnson and van Wagner, 1984; van Wagner, 1978). Such models provide insight into the composition of the vegetation and the evolution of fire life-histories (Johnson, 1978; Chapter 6).

Early analyses of disturbance history have been criticized for making unrealistic assumptions (Johnson and van Wagner, 1985; Clark, 1989). First, they assume that the environment has remained constant over the period of measurement. However, fires are much more likely under warm, dry conditions. Clark (1988) used data from fire scars and stratigraphic charcoal data, to show that past decade- and century-scale fluctuations in climate correspond to changes in the disturbance regime over the past 750 years. Studies such as these have become important for projecting the effects of climate change on potential fire intensity and frequency (Chapter 9). For example, Clark (1988) has shown that fire frequencies decreased dramatically with the onset of the 'Little Ice Age' about AD 1600, relative to the warm, dry period preceding it.

The second problem comes from estimating the probability of disturbance from mean return intervals without considering the shape of the frequency distribution around the mean. In some communities the risk of fire increases with time as dead fuels accumulate. In others, fuels, and the probability of fire, remain more or less constant as the stand ages. Clark used the frequency distribution of measured return intervals to derive a 'hazard function'. This predicts the changing likelihood of fire as a stand ages, taking this variability into account. We will return to 'hazard functions' in Chapter 6 when considering life-history evolution in fire-prone environments.

Long-term reconstruction of disturbance history in areas without long-lived trees with fire scars remains a problem. However, fire maps or written records of fires can be used to analyse short-term disturbance history. In South African fynbos shrublands, for example, fire maps and written records were used to analyse contrasting disturbance history under intensive and extensive fire management (Brown *et al.*, 1991; van Wilgen and van Hensbergen, 1992) (Figure 2.4). The intensively managed Swartboskloof area showed low probabilities of fire in young stands, attributed to prescribed burning only in mature (>10-year-old) stands. Prescribed burning had less impact on fire history in the extensively managed Cedarberg area. Brown and his co-workers were unable to distinguish between periods where fire suppression, on the one hand, and prescribed burning, on the other, were applied, a major aim of their study.

Whitney (1986) used 150-year-old survey records to reconstruct a disturbance regime for the pre-settlement pine forests of Michigan. These records included detailed accounts of tree species and of the occurrence of fire. He was able to show that fires were controlled by the nature and the

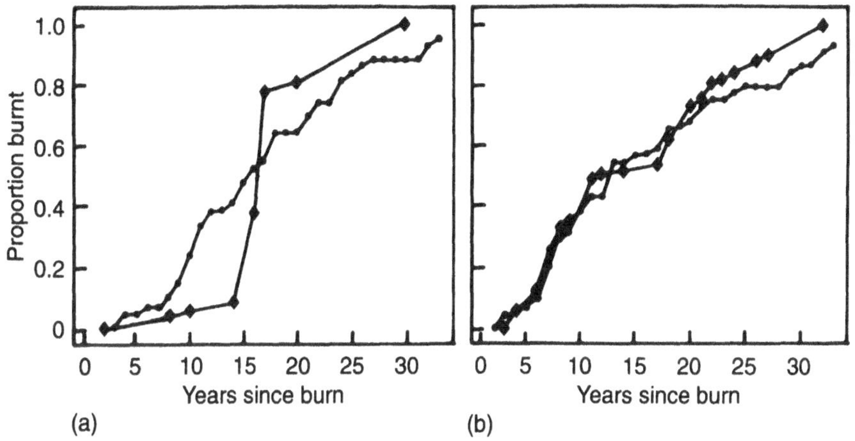

Figure 2.4 The cumulative probability of fire (lines with diamonds) with increasing age since last burn, estimated from fire maps in (a) an intensively managed area, Swartboskloof, and (b) an extensively managed area, Cedarberg, in South African fynbos. Lines with closed circles are predicted probability from a stochastic model of fire frequency (after van Wilgen and van Hensbergen, 1992). The low probability of fires in young stands in (a) reflects a successful prescribed burning programme that ensured that only older vegetation was burnt.

flammability of the vegetation, by demonstrating a close association between forest type and fire frequency.

All of the above work has contributed significantly to our understanding of fire regimes. An understanding of the fire history of a site, as opposed to simple measures such as post-fire age, is clearly important for understanding the current vegetation composition.

2.5 TEMPERATURE OF FIRES AND SURVIVAL OF PLANT TISSUES

Much of the variation in plant response to burning can be attributed to varying sensitivity of different species, or tissues, to heating (Chapter 5). A wide range of 'lethal' temperatures has been reported in the literature. The length of time required to kill plant tissue is exponentially and inversely proportional to the time of exposure to a given temperature (Figure 2.5). The moisture content of the plant material also plays a role in its susceptibility to temperature. Plants with higher moisture contents are killed at lower temperatures (Figure 2.5). Thus plants in actively growing vegetation are more easily killed than in dormant vegetation.

It would be convenient if physiological information on lethal temperatures could be coupled with temperatures generated by fires of different intensity to predict plant responses. However, the relationship between fire intensity and temperatures experienced during vegetation fires is controversial. Although many studies in fire ecology have used the temperature

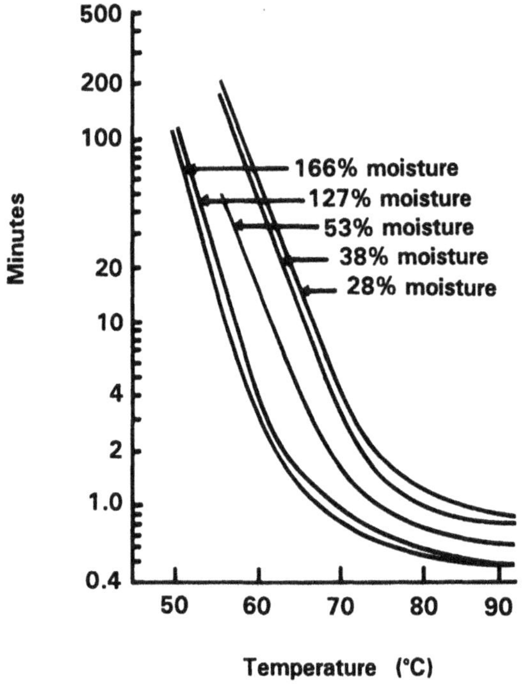

Figure 2.5 The time required to kill tissue of needle-and-thread grass (*Stipa comatu*) for different plant moisture contents in relation to temperature (after Wright and Bailey, 1982). Moisture content is measured as a percentage of dry weight.

measured during fires as an indication of fire intensity, the approach is flawed (van Wagner and Methven, 1978; Alexander, 1982). Direct temperatures, as measured during fires in the field, are not easily related to ecological effects. To quote van Wagner and Methven (1978): 'Maximum temperatures in most forest flames are about 800–1000°C and occur in a single burning pine needle as readily as in a crown fire'.

However, if used carefully, temperature measurements during fires may provide useful information that can be related to the response of the vegetation. For example, the use of heat-sensitive materials can provide information on the variability of fires, which can be related to the response of the vegetation (Hobbs *et al.*, 1984a; Hobbs and Atkins, 1988).

2.6 CONCLUSIONS

This chapter has indicated that fire is not a unitary phenomenon. The effects of fire depend on the fire regime. Different communities experience different fire regimes. A change in the fire regime can change the properties of communities (Chapters 4, 5 and 7). We have discussed the components of the fire regime and how they have been defined and measured, often by managers concerned with controlling fires. In some cases, such as fire intensity, fire regime variables defined for fire control purposes seem poor predictors of biological effects. This is a field that requires more work.

The propensity to burn depends both on the attributes of individual plant species and the collective properties of species assemblages. Plant tissues with low moisture content, especially dead material, burn most readily. Fire therefore often 'consumes' biomass which has not been consumed by herbivores. The coincidence of traits that predispose some communities to burning, while others have properties that make them resistant to fire, has led to the suggestion that flammability may have evolved (Mutch, 1970). Fire and the communities it consumes are interdependent, according to this argument, and exclusion of fire would lead to the colonization of flammable communities by more competitive non-flammable species. We return to this controversial hypothesis in Chapter 6.

3 Surviving fires – vegetative and reproductive responses

Fire directly affects plant growth, survival and reproduction and can impact the dynamics of seeds and seedlings. Fire is one of the few natural disturbances that, alone or in combination with other forces, regularly kills mature plants. This makes it an important agent in structuring communities, since the new openings created provide the potential for vegetation change. Amongst herbivores, perhaps only elephants or bark-feeding vertebrates approach fires in their importance as killing agents.

Fire has complex effects on plant reproduction. Many species are stimulated by fire, showing enhanced flowering, seed germination and seedling recruitment. Some have an obligate dependence on fire for reproduction and can be driven to extinction if fires are suppressed. To predict the ecological effects of fire we need to understand both the magnitude of fire-induced mortality, which frees up space, and the importance of fire-stimulated recruitment to utilize it.

In this chapter we discuss the immediate response of plants to fire in terms of plant survival and reproduction. These responses are used to identify four major functional types which bring some order to the many studies on fire-prone plant populations.

3.1 VEGETATIVE SURVIVAL

In the first instance, fire survival depends on the degree to which living buds are protected from heat. In woody plants, survival depends both on the extent to which the live crown is scorched and the survival of stem and root cambial tissue. The cambium is insulated from fire heat by bark. The thickness of bark, more than its type, confers most fire resistance (Figure 3.1). Small differences in bark thickness produce large differences in fire resistance, since the duration of a heat pulse required to kill the cambium of a tree is proportional to the square of the bark thickness (Hare, 1965). Figure 3.2 shows an example from experimentally heated *Jacaranda* bark.

For a given species, bark thickness increases with stem diameter, which depends primarily on age. Thus fire will tend to 'prey' selectively on younger or smaller individuals in a population, and thinner-barked species within a community (Peterson and Ryan, 1986; Uhl and Kauffman, 1990).

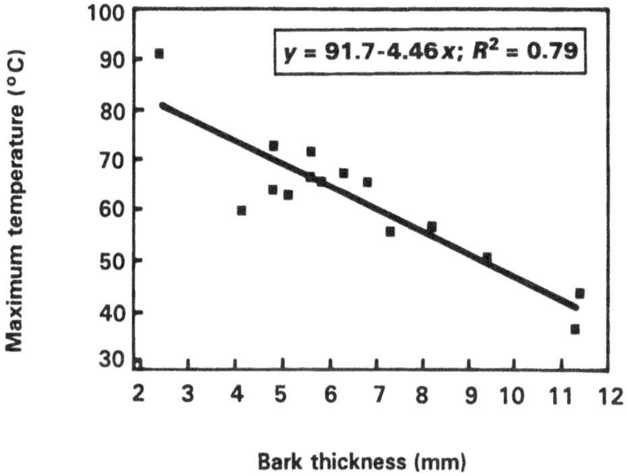

Figure 3.1 The relationship between bark thickness and maximum cambial temperature for 15 species of tropical forest trees heated in the same way; note that bark thickness accounts for most of the variation in thermal diffusivity (Uhl and Kaufmann, 1990).

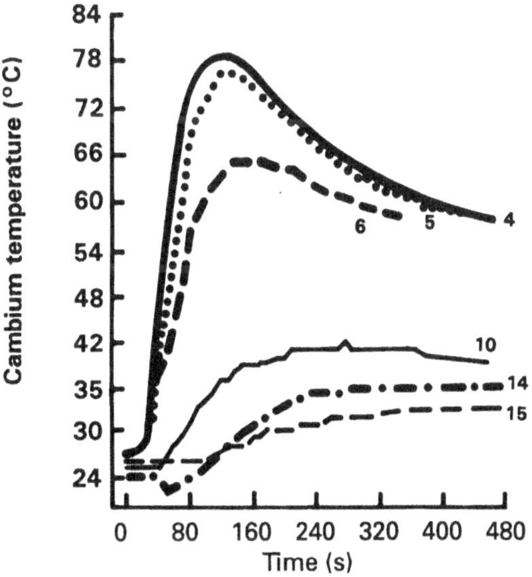

Figure 3.2 The relationship between bark thickness and cambial temperature in *Jacaranda copaia* in an experiment where the bark was heated by means of wicks. Mean flame residence time was 141 (±1.9) seconds, simulating a low-intensity surface fire. The sequence of bark thickness from highest to lowest cambial temperature was 4, 5, 6, 10, 14 and 15 mm. (Modified from Uhl and Kaufmann, 1990.)

In this way it is analogous to herbivory in selectively killing suppressed individuals (Crawley, 1983).

Canopy scorch is the second major factor influencing fire survival. In conifers, the level of crown (= canopy) scorch is the most widely used predictor of tree mortality. The level of crown scorch depends both on flame length, or vertical penetration of heat, and canopy architecture. Species with long, bare trunks and canopies held high above the ground can escape defoliation by fire (Peterson and Ryan, 1986). Growth rates influence susceptibility to crown scorch since quick-growing trees remain vulnerable for shorter periods than slow-growing trees. For example, Ponderosa pine is less sensitive than white fir (*Abies*) to defoliation by ground fires in the Sierra Nevadas because its crown rapidly grows out of reach of ground fires (Kercher and Axelrod, 1984). Fire is thus both highly selective 'predator' and defoliator in terms of size and species of its 'prey'.

3.1.1 Above-ground sprouting

The response of plants to defoliation by fire varies considerably among species (Figure 3.3). Many trees are extremely sensitive and will be killed outright if defoliation (crown scorch) is above some threshold value. Other woody plants can tolerate complete defoliation but sprout new shoots from epicormic buds, if the cambium is not damaged, thus quickly regenerating the canopy. However, even tolerant species can be eliminated by repeated defoliation from fires in short succession (Poissonet *et al.*, 1981; Zedler *et al.*, 1983).

3.1.2 Below-ground sprouting

Soil is an effective insulator since heat input into the soil is a small fraction of the total heat released by fire (Gill, 1981b). Subterranean buds are well insulated from lethal fire temperatures. The capacity to resprout from buds protected below ground is very widespread and probably a primitive condition in woody angiosperms in response to any form of severe defoliation, including fire, herbivory or frost. It is much rarer in conifers, although some, such as the California redwood (*Sequoia sempervirens*) and jack pine (*Pinus banksiana*), do resprout when burnt. Resprouting usually occurs from buds at the base of the stem, but resprouting also occurs from roots and horizontal rhizomes (Figure 3.3). Root suckering can lead to extensive vegetative proliferation after fires, producing large clonal populations, for example of birch and aspen in boreal forests (James, 1984).

In some woody plants, buds proliferate at the bases of stems and form a woody swelling known as a lignotuber or burl (Kerr, 1925; Carr *et al.*, 1984; James, 1984). Lignotubers form from a swelling in the axils of the cotyledonary leaves – they are an ontogenetic feature and not simply a response to wounding. In extreme cases, they form large, woody, plate-like structures several metres in diameter (Figure 3.3). Lignotubers occur in most *Eucalyptus* species and in at least another 11 families in Australia, Africa and North America, including Proteaceae, Ericaceae, Rosaceae and

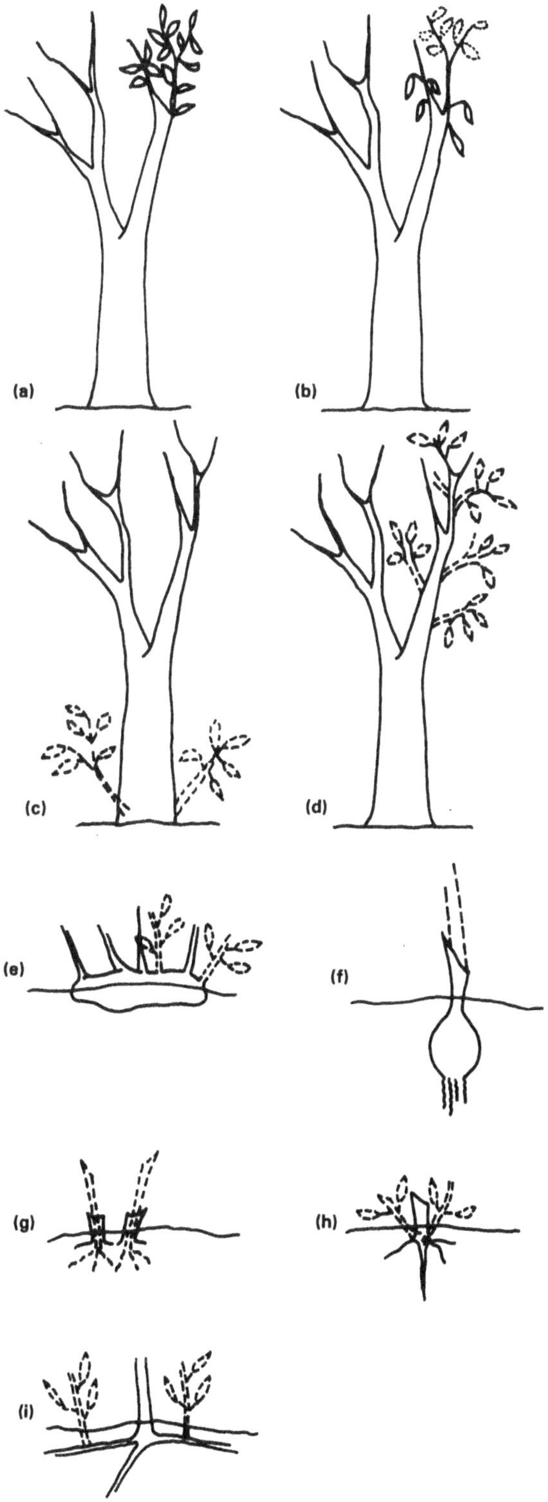

Figure 3.3 Sprouting behaviour showing both above- and below-ground responses: (a) canopy unaffected; (b) sprouting only from unaffected pre-fire buds; (c) basal stem sprouts; (d) sprouting from dormant buds insulated by bark; (e) lignotuber; (f) geophyte; (g) sprouting from grass leaves insulated by leaf sheaths; (h) root-crown sprouts; (i) root suckering.

Fabaceae (James, 1984). They are common in fire-prone shrublands but also occur as 'underground trees' ('geoxylic suffrutices'; White, 1976) in regularly burnt savannas. The function of lignotubers is still poorly understood. They are often thought to be an adaptation to fire, providing a source of buds which may sprout after injury to the aerial stem. However, the lignotubers of some species lie on the soil surface where buds are poorly insulated. They have an important function as a store of both carbohydrates and nutrients.

Monocotyledons

Grasses are among the most fire-resistant components of plant communities and, where seasonal climates lead to curing, provide highly flammable fuel. Grasses survive fire by continuing leaf growth from intercalary meristems and from new tillers growing from protected buds (Daubenmire, 1968). In tussock grasses, tiller initials are placed at or above ground level where they are protected by persistent leaf sheaths in a basal tuft. The insulating layer of old leaves protects existing leaf tissue and allows very rapid recovery by leaf regrowth. Regrowth may also occur from buds on subterranean rhizomes but is not as rapid. Grasses are most vulnerable to fire damage if they are burnt during active growth periods. Regrowth is then initiated from axillary buds and is much slower than from undamaged intercalary meristems (Booysen *et al.*, 1963; Chapter 5).

The geophytic growth form, common in many monocots and some dicots, is extremely effective for surviving fires because of insulation afforded by the soil. Some geophytes have contractile roots which pull the storage organs of young plants below the ground, protecting them from lethal fire temperatures. Arborescent monocots often rely, like grasses, on insulation from tufted crowns or persistent leaf bases. An interesting example is *Aloe ferox* from South Africa, a tall (2–4 m) single-stemmed monocot with a terminal rosette of succulent leaves. Dead leaves persist as an insulating layer on the stem (Bond, 1983). If leaves are removed, the plants suffer heavy mortality from the periodic fires in their shrubland habitat. Leaves are of medicinal value and are commercially harvested in some areas, stripping the plant of its insulating layer and causing the death of many plants when harvested populations are burnt (Table 3.1).

3.1.3 Variation in sprouting ability

As in the case of crown scorch and bark thickness, so too the capacity of species to resprout after defoliation varies with age, size and fire severity. Some species only have the capacity to sprout when they are young (Kayll and Gimingham, 1965; Hobbs and Mooney, 1985), including many leguminous savanna trees. Others show a linear increase in the capacity to resprout as they grow larger and older (Stohlgren and Rundel, 1986;

Table 3.1 Fire survival of *Aloe ferox* with and without a covering of persistent dead leaves, South Africa (Bond, 1983). Survival is compared in two populations, in one of which leaf harvesting had taken place. Leaves were collected for medicinal purposes, accounting for the high proportion of bare-stemmed individuals and much higher fire mortality in the harvested population

State of leaf layer	*Number of plants*		*% Mortality*
	survived	*died*	
In untreated population			28.8
Leaves intact	141	3	2.1
Leaves absent	5	56	91.8
In harvested population			61.0
Leaves intact	39	2	4.9
Leaves absent	0	59	100.0

Chapter 4). The capacity to sprout is not necessarily a fixed species trait and geographic variation occurs within some species. For example *Purshia tridentata*, a shrub in western American rangelands, sprouts in most populations but never sprouts in some (Driscoll, 1963).

3.1.4 Summary of survival traits

Thick bark, crown architecture and resprouting from buds protected by bark or by their subterranean position all contribute to fire survival. Fire survival varies not only among species but also among size classes, and also depends on the nature of the fire. Species with the most effective vegetative fire-survival traits have populations least influenced by fire. For species without such traits, fire-induced mortality is a key factor in regulating population growth and community composition (Chapter 4).

3.2 THE POST-BURN ENVIRONMENT AS REPRODUCTIVE STIMULUS

One of the ironies of fire ecology is that plants with thin bark, canopies consumed by fire and lacking any capacity to resprout often dominate fire-prone vegetation. Some Australian *Eucalyptus* species, for example, far from producing thick, fire-resistant bark in their fire-prone habitats, have flammable or abscising bark (Gill, 1981a). Non-sprouting species which are killed by fire and survive only from seeds come from divergent phylogenetic ancestries and different geographic origins. Typically they regenerate only after fire, and seed germination is cued to post-fire conditions. These and other species with reproduction stimulated by fire are of special interest. Their distribution and abundance is closely tied to the fire regime – they are disturbance dependent. These disturbance-dependent populations and communities are common in many parts of the world and are of great applied interest.

3.2.1 Reproduction and resource availability

What is it about fires that stimulates many species to reproduce? Post-fire conditions have many advantages for seedlings. Space is freed by the burning of established plants, resources (light, water and nutrients) increase, and seed and seedling predation declines.

The post-fire environment is characterized by high light, high temperatures, high water availability and high levels of nutrients. In addition, mineral soil is exposed where fires are hot enough to burn off litter layers. High light and temperature are obvious consequences of defoliation by fire. High water availability is a result of the reduction in transpiring leaf surface area. In chaparral, fynbos and Australian heathlands, soils may become water repellent after fire, leading to locally severe conditions of drought, especially for seedlings. For plants with deeper root systems, however, or those that penetrate the repellent layer, large reservoirs of water are available.

High nutrient availability after fire is commonly, though not universally, observed. Nitrogen loss occurs through volatilization. The more intense the fire, the greater the volatilization losses. Nitrogen volatilization from plant matter or litter, for example, begins at 200°C and increases to about 60% at 700°C (MacLean *et al.*, 1983). Volatilization losses of cations are invariably much lower (MacLean *et al.*, 1983). Despite these volatilization losses, the main effect of fire is to increase the availability of almost all nutrients. This is because nutrients are added to the surface of the soil as ash, nutrient availability is increased from heating of soil organic matter, and rates of biological mineralization increase due to changes in soil pH and reduction in C:N ratios (Stock and Lewis, 1986). The result is a nutrient 'pulse' much larger than that from, for example, rainfall or normal litterfall, for the first few months after a fire. Many researchers have reported on the dramatic increases in growth rates after fire, attributed to the high availability of nutrients and higher nitrogen concentration in the leaves (Hulbert, 1988). The post-fire environment is thus one that favours plants with high light and high nutrient requirements.

Fire also transforms the habitats of animals. The post-fire environment is hostile for many small vertebrates and invertebrates, both because of the immediate scarcity of food but also the lack of shelter and cover from predators. The post-fire decline in granivorous rodents and ants is of special significance for plant species that recruit by seed after fire. A number of studies have documented either a decline in small mammal granivores or levels of seed predation or both in the first few months after fire (Table 3.2;

Table 3.2 Changes in numbers of seed-eating, small mammals after a fire in Cape fynbos (Midgley and Clayton, 1990). Numbers represent the combined number of individuals caught at two sites at each of which a grid of 50 live traps were set out for four nights

Date	*No. captured*	*No. of species*
Pre-burn		
3 March 1986	26	6
4 March 1986	FIRE	
Post-burn		
5 March 1986	2	2
8 April 1986	0	0
1 May 1986	1	1
18 July 1986	1	1
8 September 1986	2	2
10 November 1986	12	3

Fox *et al.*, 1985; Midgley and Clayton, 1990; van Hensbergen *et al.*, 1992). In Cape fynbos, exclosure experiments have demonstrated much heavier seed predation in unburnt vegetation than in recently burnt stands (Bond, 1984; van Hensbergen *et al.*, 1992). O'Dowd and Gill (1984) reported a similar predator-free window for *Eucalyptus regnans* caused by a brief reduction in ant densities after fire and leading to increased numbers of seeds escaping to germinate. The practical importance of this predator-free window has been recognized by foresters. Daubenmire (1974) reported examples where re-seeding of conifers is only successful in the short period between the exodus of seed-eating rodents after fire and their return as the vegetation recovers.

3.3 REPRODUCTION AND FIRE

For many species, the combination of open space, increased availability of resources and temporary reduction in seed predators ('enemy-free space') is highly favourable for seedling establishment in the post-fire environment. Species exploit these conditions in diverse ways. Some are stimulated by fire and flower profusely after a burn (Plate 1). Others flower regularly between fires but accumulate seeds in seedbanks. Some plants delay dispersal of seed until they are released *en masse* when the canopy burns (Plate 2). Many have soil-stored seedbanks and use a variety of specialized cues to synchronize germination to the post-burn environment. We consider each of these in turn.

3.3.1 Fire-stimulated flowering

Fire-stimulated flowering is very common in monocotyledons (Plate 1). The fire lilies, *Cyrtanthus* spp., of South Africa are geophytes with brilliant crimson flowers (Figure 3.4). They persist vegetatively between fires but only flower in the first week or two after a burn. Fire-stimulated flowering has been reported in Orchidaceae, Iridaceae, Liliaceae (*sensu lato*), Amaryllidaceae, Poaceae and Cyperaceae, especially among geophytic growth forms and graminoids (Daubenmire, 1968; Kruger, 1983; Le Maitre and Midgley, 1992). Fire also stimulates flowering in palmoid growth forms such as *Xanthorrhoea*, the strange Australian 'tall boys', and in some resprouting shrubs such as *Telopea* in Australian heaths (Pyke, 1983) and *Leucadendron salignum* and *Colpoon* in South African fynbos (Le Maitre and Midgley, 1992). These are all growth forms that suffer little material loss in fires (e.g. geophytes, graminoids, rosette trees, lignotuberous shrubs) and can marshal existing resources into reproduction rapidly. Species with greater biomass loss in fire rely instead on persistence of seeds through the fire in seedbanks insulated by soil or canopy structures.

The association between fire and flowering varies from near obligate in small herbaceous geophytes to weakly facultative in resprouting shrubs. In most examples of fire-stimulated flowering, fire is not obligatory as a flowering stimulus. The exact nature of the stimulus varies and has been attributed to ethylene in smoke (Gill and Ingwersen, 1976 for

Figure 3.4 Fire-stimulated flowering of *Cyrtanthus contractus* in a southern African savanna 2 days after a fire (B. van Wilgen)

Xanthorrhoea) or changes in temperature, light and nutrients (Mark, 1965; Old, 1969). Keeley (1993) recently reported that flowering in *Cyrtanthus ventricosus* was stimulated by smoke but not by ethylene. The reduction in competition and increased availability of light, water and nutrients after fire is probably sufficient stimulus in many species.

Fire-stimulated flowering may have the effect of satiating pre- and post-dispersal seed predators. Mark (1965) found that seed set was greater in the tussock grass, *Chionochloa rigida*, in prolifically flowering burned plots, and seeds had a smaller chance of losses to predators, such as the larvae of gall midges. Le Maitre (1984) studied fire-stimulated flowering in *Watsonia pyramidata* in Cape fynbos. This species flowered prolifically after an autumn burn (Plate 1) but only sporadically after spring burns. He showed significantly higher levels of pre-dispersal ovule predation (mean greater than 25%) in the sparsely flowering populations than in the mass-flowering ones (mean less than 5%).

The seeds of fire-flowering species typically germinate readily. However, seeds only become available after a burn so that population growth will still be confined to the immediate post-fire period.

3.3.2 Fire-stimulated seed release

Some species accumulate seeds in a seedbank stored in the canopy of a plant, insulated from fire by cones, woody capsules or persistent woody inflorescences (Plate 2, Figure 3.5). The protective structures only open and release seeds *en masse* after fire. This condition is known as serotiny (sometimes bradyspory in the Australian literature). Serotiny differs from soil seedbanks in that the seedbank persists only for the life of the parent. Typically, serotinous species are relatively long-lived woody plants. Serotiny is very common in woody communities burnt by crown fires at intervals of a decade or more. *Eucalyptus regnans* is a remarkable example of a serotinous species that thrives on stand-destroying fires every 200–300 years (Gill, 1981b). Serotiny occurs exclusively in conifers in the northern hemisphere (*Pinus*, *Cupressus*, *Picea*, *Sequoiadendron*). In the southern hemisphere it is also found in many angiosperm families, especially in the shrubland floras of South Africa and Australia (Table 3.3). There is even a serotinous heath (*Erica sessiliflora*) in the Cape fynbos. Serotinous species dominate many of the communities in which they occur, and there is increasing interest in their biology (see Lamont *et al.*, 1991b for a recent review). Serotiny is not unique to fire environments but is common in some arid land plants where good rains, and not fire, trigger seed release (Gunster, 1992).

The degree of serotiny varies greatly both between and within species. Cones may remain closed for little more than a year in some species to more than 30 years in others (Vogl *et al.*, 1977). In Bishop pine, *Pinus attenuata*, cones have been found completely enclosed by wood but have yielded

(a)

(b)

(c)

Figure 3.5 Examples of serotiny, showing pre-burn condition on the left and post-burn condition on the right. (a) Woody cone, *Widdringtonia* (Cupressaceae); (b) woody follicle, *Hakea* (Proteaceae); (c) cone formed by inflorescence bracts, *Protea* (Proteaceae).

viable seed when opened. The mechanism of seed release is very variable. In some species of pine, waxes and resins seal the cone scales and seeds are only released when the waxes are melted by fire heat. Fire is obligatory for seed release. In Australian *Banksia*, seeds are enclosed between two woody valves in a follicle. In *Banksia ornata*, the valves rupture under the combined influence of high fire temperatures and low follicle moisture content

Table 3.3 Taxonomic and geographic distribution of serotiny. Approximate fire frequencies refer to crown fires. With the possible exception of *Staberoha* and *Thamnochortus* (Restionaceae; Bond, personal observation), no serotinous monocots have been reported. (From Kruger, 1977; Vogl *et al.*, 1977; Rowe, 1983; Bond, 1985; Lamont *et al.*, 1991)

Taxon	*Example*	*Region*	*Vegetation*	*Fire frequency* (y)
Gymnosperms				
Pinaceae	*Pinus attenuata*	North America	Chaparral	50+
	P. banksiana	North America	Boreal forest	50+
	P. contorta	North America	Conifer forest	50+
	P. rigida	North America	Shrublands, forest	15+
	P. halepensis	Mediterranean	Matorral, woodland	50+
	Picea mariana	North America	Boreal forest	50+
Cupressaceae	*Cupressus*	North America	Chaparral	50+
	Widdringtonia	Africa	Fynbos	15+
	Actinostrobus	Australia	Heathland	15+
	Callitris	Australia	Heathland, woodland	15+
Taxodiaceae	*Sequoiadendron*	North America	Conifer forests	100+ (crown fires)
Angiosperms				
Asteraceae	*Helipterum*	Africa	Fynbos	15+
	Phaenocoma	Africa	Fynbos	15+
	Oedera	Africa	Fynbos	15+
Bruniaceae	*Brunia*	Africa	Fynbos	15+
	Berzelia	Africa	Fynbos	15+
	Nebelia	Africa	Fynbos	15+
Casuarinaceae	*Casuarina*	Australia	Woodland	15+
Ericaceae	*Erica sessiliflora*	Africa	Fynbos	15+
Myrtaceae	*Eucalyptus*	Australia	Woodland	15–200+
	Callistemon	Australia	Shrubland, woodland	15+
	Leptospermum and others	Australia	Shrubland, woodland	15+
Proteaceae	*Aulax*	Africa	Fynbos	15+
	Banksia	Australia	Shrubland, woodland	15+
	Dryandra	Australia	Shrubland, woodland	15+
	Hakea	Australia	Shrubland, woodland	15+
	Leucadendron	Africa	Fynbos	15+
	Protea	Africa	Fynbos	15+
	and others	Australia	Shrubland, woodland	15+

(Gill, 1976). The moisture content of the follicles declines with age or the death of parental branches, resulting in some seed release into unburnt vegetation. The time taken for valves to rupture decreases with increasing temperature (Gill, 1976). The more strongly serotinous *Banksia* species provide their own fuel bed around the follicle in the form of persistent florets. These burn, providing the necessary high temperature cue needed to rupture the follicles (Lamont and Cowling, 1984). *Banksia* species also exercise control on the timing of seed release by possessing a 'separator' which scoops seeds out of the follicles by hygroscopic movements when exposed to the wet–dry cycles of the rainy season (Cowling and Lamont, 1985b). In fynbos Proteaceae, cones seldom persist for more than 6 or 7 years on a bush. Consequently, there is a small but annual supply of seeds falling from old cones, although major seed release only occurs after fire.

3.3.3 Fire-stimulated seed germination

Many species with soil-stored seedbanks have fire-stimulated germination. There is a burgeoning literature on the nature of fire-stimulated germination cues (e.g. Went *et al.*, 1952; Keeley, 1991; Bell *et al.*, 1993). These can have an important bearing on the ecology of populations exposed to different fire regimes (Chapter 5). For example, heat-stimulated germination of legume seeds can fail if fires are of low intensity. In Western Australia, buried seeds of *Acacia* trees failed to germinate in low-intensity management burns in jarrah (*Eucalyptus marginata*) forests. The resulting low density of *Acacia* shrubs not only reduced nitrogen input into the system but also led to the spread of the pathogenic fungus, *Phytophthora cinnamomi*, which causes dieback of the valuable jarrah trees (Shea *et al.*, 1979). A precise understanding of the 'germination niche' provides a useful predictive tool for fire management of this system and others like it (Chapter 5).

Physical cues

Seeds are cued by a variety of signals associated with fire but the most common appear to be the heat shock of the fire itself or chemicals released in smoke or charred wood. The most well-studied case of heat-induced germination is 'hard-seededness', where dormancy is due to an impermeable seed coat (Quinlivan, 1971). Dormancy in legumes is broken by rupturing of the lens, a specialized part of the seed coat. The opening of the lens in *Acacia* is induced by heat and effected through the splitting of thin-walled cells within the lens (Cavanagh, 1980). After the lens is lifted, water enters the seed via the lens cavity and germination can begin. Hard-seededness also occurs in other families, including Rhamnaceae, Cistaceae, Convolvulaceae, Geraniaceae, Malvaceae and Proteaceae (Gill, 1981b;

Keeley, 1991; Thanos *et al.*, 1994). The mechanism of dormancy breaking is less well known in these groups and may be quite different. For example, heat causes desiccation of the seed coat in a fynbos *Leucospermum* species which then cracks on moistening, initiating germination (Brits *et al.*, 1993). Dormancy of hard seeds may be broken not only by fire but also by scarification in stream beds, passage through vertebrate guts or by cracking of the seed coat caused by trampling of the soil, so that the trait cannot be considered a unique adaptation to fire.

As with most forms of seed dormancy, hard-seededness is quite variable within a species or population and some seeds germinate without a heat stimulus. Nevertheless, the heat associated with fire is important in the breaking of seed dormancy in many fire-prone systems.

Chemical cues

Fire alters the chemistry of the environment and these chemical changes can cue fire-stimulated germination. Californian chaparral fire ephemerals provide an interesting example. Chaparral is characterized by a dense shrub cover with a sparse herbaceous understorey. The virtual absence of herbs under mature chaparral was first attributed to allelopathic inhibition of germination by leachate from the shrub canopy (McPherson and Muller, 1969). Fires would remove the source of the allelopathic chemicals, thus triggering the characteristic luxuriant herb growth after fire. Subsequent studies have tested this and a variety of alternative hypotheses, including unfavourable abiotic conditions and high levels of herbivory beneath the shrub canopy (e.g. Christensen and Muller, 1975). Taken together, these studies showed that mature chaparral was a very hostile habitat for an herbaceous seedling. Far from being inhibited by allelopathic toxins, seeds would be better off delaying germination until a fire created a more favourable environment. However, there was no evidence for hard-seededness in most of the fire-annuals.

Wicklow (1977) was the first to discover that seeds of a common fireweed, *Emmenanthe penduliflora*, were stimulated to germinate in the presence of charred wood. Subsequent work by S. Keeley and J. Keeley (Keeley *et al.*, 1985; Keeley, 1991) has shown that many fire-annuals, a number of subshrubs and a few shrub species, including the widespread chamise *Adenostoma fasciculatum*, have seed germination stimulated by charred wood. An example of experimental results is given in Table 3.4. The active principle in charred wood extracts – probably an oligosaccharin derived from a hemicellulose compound – is effective at low concentrations (S. Keeley and Pizzorno, 1986). The nature and mode of action of the charred wood stimulant is still unknown. Ashed wood has no effect. The charred wood cue, directly related to the incidence of fire, has since been found in a few species in fire-prone shrublands of South Africa and Australia, and the Mediterranean basin (Keeley, personal communication, 1991).

Table 3.4 Germination (%) of chaparral plants in response to a heat shock and the presence of charred wood (Keeley, 1991). *Ceanothus* is a typical hard-seeded species requiring a heat shock. *Garrya*, *Romneya* and *Phacelia* are not hard-seeded and germination is stimulated by charate

Temperature (°C) *Time (min)*	*Control*	*70* *120*	*80* *60*	*100* *5*	*120* *5*
Ceanothus megacarpus (shrub)					
Control	11	41	–	48	80
Charred wood	2	40	–	40	61
Garrya flavescens (shrub)					
Control	4	12	–	10	6
Charred wood	65	34	–	54	19
Romneya trichocalyx (suffr.)					
Control	0	0	–	0	0
Charred wood	24	33	–	34	24
Phacelia cicutaria (annual)					
Control	5	–	11	–	6
Charred wood	32	–	45	–	57

Recent studies have also shown that seed germination can be stimulated directly by smoke. Smoke stimulation was first reported for *Audouinia* (Bruniaceae), a rare shrub species in the Cape fynbos (de Lange and Boucher, 1990). It has since been reported in many other fynbos species (Brown, 1993), in ecologically similar Australian heathlands (Dixon *et al.*, 1995) and even in the grass species, *Themeda triandra*, one of the dominant components of southern African grasslands (Baxter *et al.*, 1994). The physiological mechanism underlying smoke-stimulated germination is not yet known and may vary from species to species. Though smoke and charate stimulation suggest obligate reproductive dependence on fire, not all seeds, or populations, respond to these cues. It is not yet clear whether the chemicals involved are general germination stimulants or active only in fire-prone species.

As in the case of fire-stimulated flowering, many plants which germinate after fires probably use generalized gap-responding cues. For example, germination of the small seeds of *Calluna vulgaris* is stimulated by brief exposure to light and fluctuating temperatures, both characteristic of the post-burn environment but also of other gap-creating disturbances. *Calluna* germination is also slightly stimulated by the fire heat pulse (Whittaker and Gimingham, 1962). Nevertheless, the demographic effect of fire-stimulated flowering, seed release or germination is similar in tending to produce even-aged cohorts each dating from the last fire.

3.4 FIRE LIFE-HISTORIES

Plant life histories are measured in terms of life span and frequency of reproduction, producing categories such as annuals, pauciennials, iteroparous and semelparous perennials. Where fire is important, it is useful to change the unit of time from years to fire intervals marking the passage of generations. For example, fire-recruiting non-sprouting species are replaced after each fire by a new generation of seedlings, so the generations are non-overlapping. Other species, the fire ephemerals, also emerge only after fires but are short-lived and do not survive vegetatively until the next fire. Sprouting species have overlapping generations but some recruit seedlings only after a burn, whereas others recruit in older vegetation between burns.

There have been several attempts to draw up a classification of species into fire life-history categories. Gill's (1981a) widely used classification is shown in Table 3.5. This classification emphasizes the mode of fire survival (sprouting/non-sprouting) and the nature and location of the regenerating tissue (seeds/buds, above ground/below ground). Gill does not emphasize the timing of reproduction, although it is also useful to know whether, for example, seedling establishment is stimulated by fire. Keeley (1986) takes this into account by defining species as fire-recruiters or fire-resisters. Fire-recruiters show reproduction stimulated by fire while fire-resisters typically survive fires (usually by sprouting) but recruit seedlings only in older, unburnt vegetation. Unfortunately this binary classification confounds vegetative and reproductive responses to burning.

Table 3.5 A classification of plant species in relation to their response to fires (ex Gill, 1981a). The table can be used as a key. Note that this system does not incorporate seedbank properties of resprouters (see text)

A Annuals, pauciennials (life history not linked to fire)

B Fire ephemerals (appear soon after fire, short life-span and usually dead before next fire, depend on fire for regeneration)

C Non-sprouting (reproductively mature plants with 100% leaf scorch die)
- (a) Seed stored on plant (serotinous)
- (b) Seed stored in soil
- (c) No seed storage (dependent on dispersal)

D Sprouting (reproductively mature plants with 100% leaf scorch survive)
- (a) Buds below ground
  - (i) root suckers or horizontal rhizomes
  - (ii) basal stem sprouts or vertical rhizomes
- (b) Buds above ground
  - (i) epicormic buds
  - (ii) undamaged active pre-fire buds

Table 3.6 Fire life histories based on vegetative survival and the effect of fire on seedling recruitment

Recruitment	*Vegetative response*	
	Survive fire by sprouting	*Killed by fire, non-sprouters*
Stimulated by fire	Fire-recruiting sprouters	Fire-recruiting non-sprouters
Not stimulated by fire	Non-fire-recruiting sprouters	Non-fire-recruiting non-sprouters

We have preferred to consider survival and reproductive responses to burning independently. This produces four major fire life-histories based on the dichotomy between fire-recruiting and non-fire-recruiting reproductive timing and sprouting and non-sprouting survival modes (Table 3.6). The four groups have distinctly different population responses to fire (Chapters 4 and 5), have evolved under differing selective pressures (Chapter 6), make up different proportions of plant communities and appear at different stages in post-fire succession (Chapter 7).

The prevalence of the different life histories changes from one geographic region to another and is one reason why it is so difficult to make generalizations about the role of fire in ecosystems. Long-lived, non-sprouting species are most common in shrublands and conifer and eucalypt woodlands with fire intervals greater than 10 years. Sprouting species occur in all vegetation types. Species with non-fire-recruiting reproductive timing predominate in grasslands and savannas with relatively short fire frequencies, while fire-recruiting species are characteristic of woody vegetation with longer fire cycles. However, studies on fire responses are very unevenly distributed and it is difficult to get a global view of geographic patterns from the literature.

3.5 CONCLUSIONS

There is still a great deal to learn about the biology of plants and fire, as the recent discoveries of charate- and smoke-stimulated germination make clear. Our ignorance of the 'natural history' of fire stems partly from the prejudiced view that fires cause damage to ecosystems rather than renewal. In the Cape flora, for example, the leading botanists in the early decades of the twentieth century were so concerned with the scourge of fire that they failed to observe such obvious phenomena as fire-induced flowering, fire – stimulated seed germination and serotiny – all clear indicators of the essential role of fire in the life history of fynbos plants. One result of ignoring the natural history of fire was many years of futile attempts at fire suppression at the expense of declining populations of fire-dependent species.

Fire clearly has diverse and large effects on the survival and reproduction of many plant species. In the following chapters we address the importance of fire as an agent of ecological change arising from the diversity of fire regimes and the diversity of species responses.

4 Plant demography and fire I. Interval-dependent effects

4.1 INTRODUCTION

Plant populations respond in remarkably diverse ways to fire. Some remain essentially stable from one fire to the next. Others may fluctuate from explosive growth to local extinction (Figure 4.1). What accounts for stability in some populations and variability in others? How is fire implicated in regulating population size? Part of the variability among species responses can be explained by different fire life-histories – sprouting species should have more stable populations than non-sprouting species; non-sprouting species with persistent soil seedbanks should be more stable than, say, serotinous species with seedbanks exhausted after fire.

In this chapter and the next we evaluate three competing explanations for population changes in fire environments. For convenience we refer to them as the **fire-interval** hypothesis, the **self-regulatory** hypothesis and the **event-dependent** hypothesis (relating to the complex of events occurring post-burn). The fire-interval hypothesis argues that variation in the fire return interval is the chief cause of population variation. Fire interrupts the normal pattern of growth, mortality and reproduction of plants. The duration of these normal growth phases determines the state of the population when it is burnt and its post-fire response. The hypothesis predicts that constant fire intervals will produce constant population growth rates.

The self-regulatory hypothesis argues that population size is determined by density-dependent feedbacks regulating population size. Density dependence is one of the least studied aspects of population regulation in fire environments. At least in theory, it can produce a diversity of population responses, including stable population sizes or chaotic oscillations (May, 1974; Berryman, 1991), which mimic the diversity produced by variable fire frequencies. The study of density-dependent oscillations in plant populations has barely begun and then largely with annuals (e.g. Symonides *et al.*, 1986; Thrall *et al.*, 1989). However, the life histories of annuals have much in common with non-sprouting fire life-histories and these, too, may show oscillatory behaviour. These oscillations may be difficult to distinguish from fluctuations ascribed to extrinsic factors such as climate or fire regime. The self-regulatory hypothesis predicts that population size may fluctuate greatly even under constant environmental conditions and disturbance regimes.

The event-dependent hypothesis argues that population fluctuations are driven by the unique circumstances of a fire affecting the number of seeds that germinate or plants that survive. The killing power of a particular fire and the conditions that follow it are the issue rather than the 'normal' processes occurring in the periods between fires. The event-dependent hypothesis predicts that constant fire intervals may produce very different population effects depending on the disturbance event.

The three alternatives are not necessarily independent, nor are they mutually exclusive. Long return intervals will allow more time for density-dependent phenomena to be expressed. Longer periods for fuel accumulation may result in more intense fires and so alter the impact of the fire event on survival and reproduction (Chapter 5).

We first discuss the implications of the four different fire life-histories

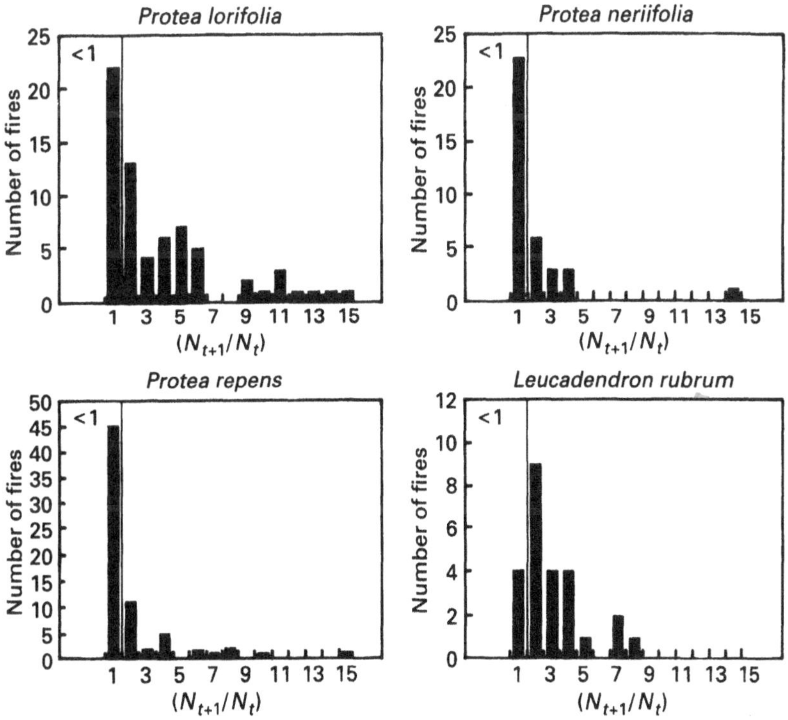

Figure 4.1 Variation in net reproductive rates (N_{t+1}/N_t) after fires for serotinous, non-sprouting species of fynbos Proteaceae. Rates were censused from post-burn surveys of seedlings, N_{t+1}, and the skeletons of burnt parents, N_t (Bond, 1989). Class intervals are 1 = 0–0.99, 2 = 1–1.99, 3 = 2–2.99, etc. The area left of the vertical line indicates recruitment below replacement levels ($N_{t+1}/N_t < 1$).

introduced in Chapter 3 for demographic processes. We then develop some simple population models appropriate for a fire-prone environment. Finally, we consider data and models that address the different hypotheses for regulatory controls on plant populations. Many of our examples are drawn from studies of trees and shrubs from Mediterranean-type shrublands. These provide easily understood and well-studied examples of plant demography and fire. However, we also discuss other growth forms, such as grass and savanna trees. This chapter, together with the next, provides an overview of the diverse, but important role of fire in regulating plant populations.

4.2 TYPES OF POPULATION GROWTH

The four fire life-histories described in Chapter 3 produce four distinct patterns of population growth (Figure 4.2):

- Non-fire-recruiting sprouters have populations that grow between fires. Population growth is continuous with overlapping generations because of interfire seedling recruitment. Fires temporarily inhibit recruitment,

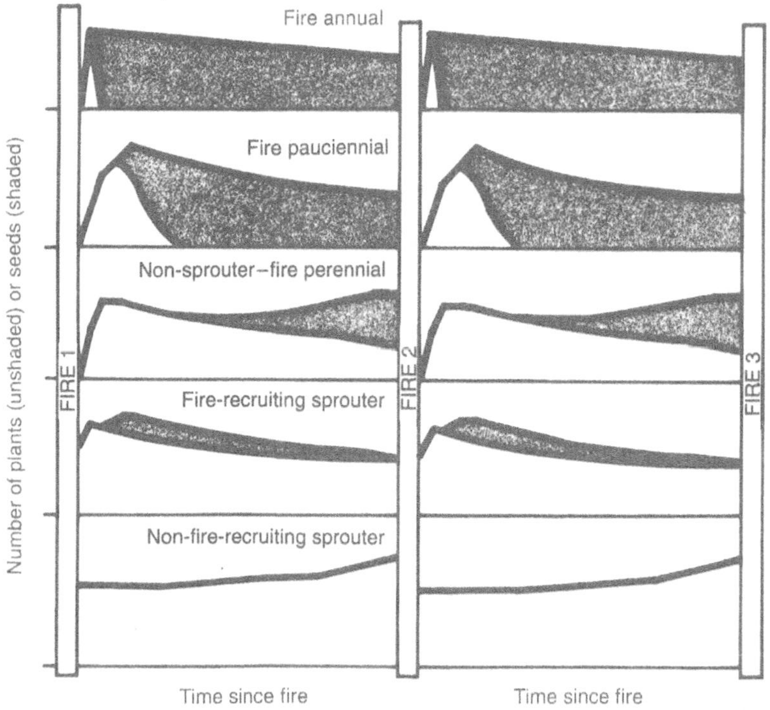

Figure 4.2 Some alternative life histories of plants exposed to fire. The unshaded area represents the number of growing plants. The shaded area represents the number of seeds, including seedbanks. Two successive fire cycles are shown.

and may increase both juvenile and adult mortality. Short fire return intervals often eliminate the fire-sensitive juveniles. Typical examples are forest tree seedlings colonizing fire-prone savanna, shrublands or grasslands (Trapnell, 1959; Keeley, 1986; Gimingham, 1988; Manders, 1990).

- Non-fire-recruiting non-sprouters. Species in this group are comparatively rare in fire-prone vegetation. They include short-lived species with well-dispersed seeds which exploit small gaps not created by fire. Longer-lived species in this group sometimes rely on local refugia from burning for their persistence. Often they are transient colonists invading fire-prone areas from neighbouring less-flammable communities. Given sufficient time, such species can transform the fuel properties of a community, causing major system shifts (Chapter 7).
- Fire-recruiting sprouters have discrete population growth with a size structure of multiple cohorts each dating from a previous fire. A typical example is chamise, *Adenostoma fasciculatum*, a resprouting chaparral shrub (Stohlgren and Rundel, 1986). Populations decline between fires in this group.
- Fire-recruiting non-sprouters include many shrubs, trees and some monocots in shrublands and woodlands. Populations form even-aged cohorts with non-overlapping generations dating from the last burn. Numbers decline between fires. The category includes fire ephemerals which emerge after fire, persist for a few years, then survive only as seeds to the next fire. In this group, a key concern is the size of the seedbank at the time of the fire. Fires can eliminate populations that have not yet accumulated sufficient seeds or those whose seedbanks are exhausted after long intervals without fire.

4.3 MODELLING POPULATION GROWTH

There is a growing number of plant demographic studies in fire environments, particularly of fire-recruiting shrubs and trees. However, fire ecologists have mostly avoided any attempt to formulate their data into models. Models are useful tools for analysing the causes of population change. Here we introduce commonly used models which can be adapted for fire life-histories. A convenient verbal model of population growth (Begon *et al.*, 1990) is:

$$N(\text{now}) = N(\text{then}) + \textbf{B}\text{irths} - \textbf{D}\text{eaths} + \textbf{I}\text{mmigrants} - \textbf{E}\text{migrants}. \qquad (4.1)$$

In practice the number of immigrants and emigrants is often ignored, especially in models of plant population growth.

4.3.1 Populations without age structure

In populations of fire-recruiting non-sprouters, the old generation is killed by fire and replaced by the new generation of surviving seedlings. Among woody species, the two generations can often be censused simultaneously by counting seedlings ($N_{t+1} = N_{now}$) and the skeletons of burnt parents ($N_t = N_{then}$). The time elapsed between 'now', $t+1$, and 'then', t, in this case covers the life of a whole generation and is the length of the fire interval. Population growth can be modelled as :

$$N_{t+1} = \lambda N_t \tag{4.2}$$

where $\lambda = N_{t+1}/N_t$, the net reproductive rate (the number of surviving seedlings per parent) and the subscript denotes the time at which N is measured. For example, if a non-sprouting shrub has an average of three surviving seedlings for every pre-burn parent, $\lambda = 3/1 = 3$. The population will triple in size.

Population size after t fires is given by:

$$N_t = \lambda^t N_0. \tag{4.3}$$

This is the equation for **geometric** growth. In its continuous form it describes exponential growth. Geometric growth is seldom sustained for long periods. After 10 fires, for the above example, an initial population of two would have grown to 118098.

In real populations, λ is never constant over growth cycles. Seedling numbers may vary for a diversity of reasons – differences from fire to fire in seedbank size, depredations of herbivores, drought deaths or changing competitive milieu. Taking account of these, λ becomes a variable, and our equation is rewritten:

$$N_{t+1} = N_t \lambda f(\text{climate, predators, competitors, etc.}). \tag{4.4}$$

Many of the factors affecting λ may be unknown or unknowable (such as next year's rainfall). It is then useful to treat λ as a random variable with mean and variance from which stochastic population growth can be simulated. Figure 4.3 shows an example using a lognormal frequency distribution of λ derived from post-fire censuses of fynbos Proteaceae illustrated in Figure 4.1. Note that the mean and frequency distribution of λ in this example are derived from spatial averaging of pre- and post-burn censuses from many fires rather than temporal averaging from the same population subjected to repeated fires. The latter would require several centuries of observations. In the particular simulation shown in Figure 4.3, there is a striking contrast between the geometric growth equation using a constant or variable λ though the mean λ is 2.5 in both cases. For most simulations, however, the stochastic growth model still approximates exponential growth and populations explode.

Populations are always bounded by available habitat. A realistic population growth model requires a 'ceiling' to cap further growth. The most commonly modelled departure from constant λ is:

$$N_{t+1} = N_t \lambda f(N_t) \tag{4.5}$$

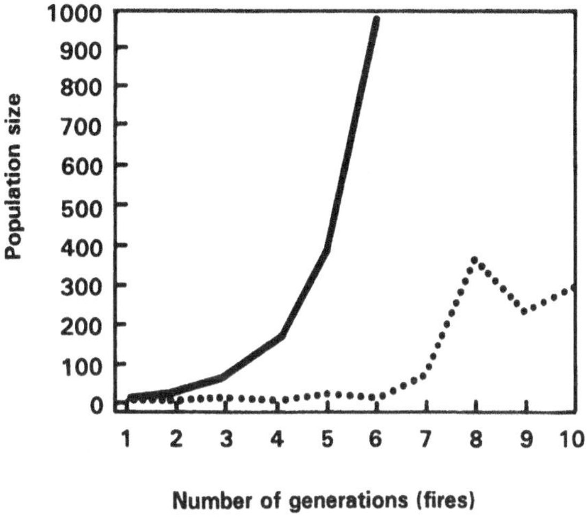

Figure 4.3 Comparison of geometric growth rates with λ constant = 2.5 (solid line) or simulated from a variable λ (dotted line) with mean = 2.5 and a frequency distribution approximating a negative exponential. A single run of the stochastic model is shown simulating growth in serotinous proteoid shrubs.

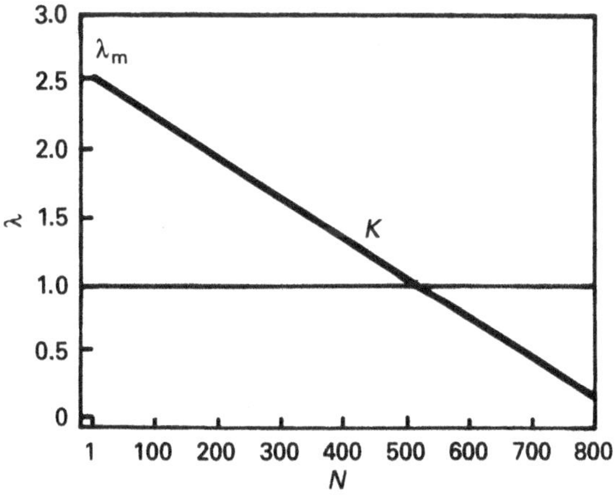

Figure 4.4 Derivation of the logistic equation. The net reproductive rate, λ, is linearly proportional to population density, N. $\lambda = N_{t+1}/N_t$, λ_m = maximum net reproductive rate, K = carrying capacity, the population size at which net reproductive rate, λ, = 1.

so that λ is treated as a variable changing with population size and diminishing as the population outstrips resources.

How would our equation change if λ was not constant but a variable dependent on N? Figure 4.4 shows one case where λ declines linearly as N increases. λ_m, the maximum population growth rate, can be read as the intercept on the Y axis and K as the intercept on the X axis where

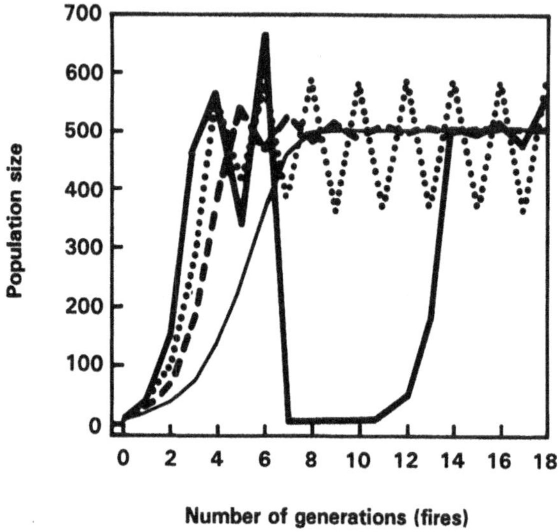

Figure 4.5 Complex behaviour produced by the discrete logistic equation. Parameters are set as $\lambda_m = 2.0$ (thin line), 2.75 (dashed line), 3.25 (dotted line) and 4 (thick line), and $K = 500$.

Y (i.e. λ) = 1. K is the **carrying capacity**, the 'ceiling' value of N at which $\lambda = 1$ with no change in population size.

Using the equation for a straight line,

$$\lambda = \lambda_m + N_t \frac{(1-\lambda_m)}{(K-0)} = \lambda_m - N_t \frac{1-\lambda_m}{K}$$

and, incorporating this new definition of λ into equation 4.5, N_{t+1} now becomes

$$N_{t+1} = N_t \left(\lambda_m - N_t \frac{1-\lambda_m}{K} \right) \tag{4.6}$$

This is a discrete form of the **logistic** equation. This simple equation produces surprisingly complex behaviour depending on the value of λ_m and K (Figure 4.5). At low values of λ_m the population grows asymptotically to K. As λ_m increases there are damped stable oscillations around K. With a further increase, stable oscillations are produced. With high values of λ_m, there are unstable oscillations of increasing amplitude (**chaos**).

There are some interesting conclusions to be drawn from this simple model of density-dependent growth:

- First, density dependence, as expected, does prevent a population from 'exploding'.
- Secondly, under a rather restricted set of circumstances, density dependence causes the population size to stabilize at a fixed ceiling from fire to fire.
- Thirdly, density dependence is not necessarily stabilizing. Populations with high reproductive potential can overshoot the steady-state size after

which density-dependent effects on birth and survivorship cause a population crash.

So-called 'chaotic' population growth (large values of λ_m) can strongly resemble stochastic population growth although the equation is deterministic. An important property of deterministic chaos is that very small differences in initial values (i.e. N_0) lead to very large differences in population size in later generations (May, 1974; Berryman and Milstein, 1989). This means that prediction of future population growth will become increasingly inaccurate over more fires.

The behaviour of density-dependent models is strongly affected by the nature of the relationship between λ and *N*. For example, if the relationship shown in Figure 4.4 was not linear but a convex curve, population growth would be more stable and chaos less likely. The inclusion of dispersal in the logistic model also tends to stabilize population growth.

Nevertheless our simple models of population growth suggest at least two explanations for natural variability in population size: stochastic variation due to environmental variation versus deterministic variation due to strong density dependence. We will discuss evidence to support both cases from demographic studies on fire-prone plants below.

4.3.2 Populations with age structure

So far we have assumed that fire frequency is more or less constant, or, if variable, that the variability is only a minor contributor to λ. Proper consideration of the effects of different fire frequencies requires information on age-specific survivorship and seed production. Take, for example, a non-sprouting shrub with a cohort of seedlings establishing only after fire. Annual seed production can be modelled as:

$$S_x = \sum_{x=j}^{m} l_x \cdot b_x \tag{4.7}$$

where S_x is the mean number of seeds per plant aged *x*; *j*, age at first seed production; *m*, maximum reproductive age; l_x, the probability of an individual surviving to age *x*; and b_x, the mean number of seeds produced per individual between age *x* and $x - 1$. If the seeds persist to form a seedbank, then both past seed production and the number surviving to the present must be calculated. Thus:

$$S_t = \sum_{x=j}^{m} l_x \cdot b_x \cdot S_{tx} \tag{4.8}$$

where S_t is the number of seeds at population age *t* and S_{tx} is the fraction of seeds produced at age *x* which survive to age *t*. Here the product $l_x b_x$ gives the number of seeds produced per individual aged *x* and S_{tx} defines their subsequent survival (Auld, 1987; Parker and Kelly, 1989). This model allows us to predict the number of seeds available for any particular fire frequency – given the necessary information on age-specific survival and reproduction and seed survival rates. Examples of this kind of analysis are discussed further below.

4.3.3 Transition matrices and stage structured models

The final model we shall consider incorporates information on survival, growth and fecundity for plant populations with a mixed age (or **stage,** e.g. seedling, immature, adult) structure. These are appropriate where fires do not kill all individuals (sprouters) or where recruitment occurs between fires so that mixed age cohorts coexist. These models are simple to use on micro-computers, easy to set up on widely available spreadsheet software and of wide applicability. They have rarely been used for modelling populations exposed to fire. A study modelling the effects of fire on a South American savanna grass, *Andropogon semiberbis* (Silva *et al.*, 1991) serves as an example.

To construct a matrix model, a **column vector** is first set up with the elements representing the number of individuals censused in each relevant age or stage class. Silva *et al.* used four stages: 1, 1 tiller; 2, 2–10 tillers; 3, 11–20 tillers; 4, >20 tillers. The probability of changing from one stage to each of the others is calculated by observing plants in the field for one or more time intervals. For example, 100 one-tillered plants might yield, in the following year, eight surviving plants still with only one tiller, 20 plants which have grown to stage 2 and no plants which grew into stage 3 or 4. Transition probabilities from stage 1 to each of the four stages would then be 0.08, 0.20, 0.0 and 0.0. Transition probabilities for each stage in the plants' life cycle are entered into a transition matrix. Reproduction is represented in the first row. Thus, from Table 4.1, stage 4 plants produce, on average, 4.4 new one-tillered offspring per year. The smallest stage class adds no new offspring but a proportion (0.08) survive as single-tillered individuals. The matrix is square with the same number of rows and columns representing the probability of every transition from one stage class to another. An example from Silva *et al.* is shown in Table 4.1.

Growth of the whole population and of each of the stage classes is calculated by multiplying the column matrix **B** of the initial numbers in each stage by the transition matrix, **A**, using matrix multiplication. For a three stage model this is calculated as:

$$\mathbf{A} \times \mathbf{B}_1 = \mathbf{B}_2 \tag{4.9}$$

$$\begin{bmatrix} a_{11} & a_{12} & a_{13} \\ a_{21} & a_{22} & a_{23} \\ a_{31} & a_{32} & a_{33} \end{bmatrix} \times \begin{bmatrix} N_1 \\ N_2 \\ N_3 \end{bmatrix} = \begin{bmatrix} a_{11} \times N_1 + a_{12} \times N_2 + a_{13} \times N_3 \\ a_{21} \times N_1 + a_{22} \times N_2 + a_{23} \times N_3 \\ a_{31} \times N_1 + a_{32} \times N_2 + a_{33} \times N_3 \end{bmatrix}$$

Table 4.1 A transition matrix model for *Andropogon semiberbis* in Venezuelan savanna grassland (ex Silva *et al.*, 1991). The stages represent: 1, 1 tiller; 2, 2–10 tillers; 3, 11–20 tillers; 4, >20 tillers. Transition probabilities are for a burnt population over 1 year's growth. Population projections from matrix multiplication are shown on the right-hand side. The stage structure stabilizes after 10 years, giving constant proportions in each age class, P. The total projected tiller population is shown for each time interval and λ is calculated from these totals for $t = 11$

		Transition matrix					Population projections				
		From									
		1	2	3	4		t_0	t_1	t_{10}	t_{11}	P
To	1	0.08	1.63	2.42	4.4	N_1	2	17.1	96.3	120.5	0.63
	2	0.21	0.64	0.35	0.16	N_2	2	2.7	40.4	50.6	0.27
	3	0.0	0.19	0.43	0.23	N_3	2	1.7	10.7	13.4	0.07
	4	0.0	0.03	0.23	0.48	N_4	2	1.5	4.8	6.0	0.03
						Total	8	23.0	152.2	190.5	
						$\lambda = N_{t+1}/N_t$			190.5/152.2 = 1.252		

With repeated multiplication of the transition matrix:

$$\mathbf{A} \times \mathbf{B}_1 = \mathbf{B}_2, \ \mathbf{A} \times \mathbf{B}_2 = \mathbf{B}_3, \ \mathbf{A} \times \mathbf{B}_3 = \mathbf{B}_4, \ldots$$

the age or stage structure of the population will stabilize at a constant ratio of age classes. This stable age or stage structure is independent of the initial numbers in each class and depends only on the coefficients of the transition matrix. When the stable age structure has been reached, λ can be calculated, as before, by N_{t+1}/N_t. An example of the calculations is shown for Silva *et al.*'s savanna grass in Table 4.1. Note that λ is greater than 1 so that this transition matrix implies the population is growing. Note too that the model includes no density dependence so growth is unconstrained and the population will eventually 'explode'.

It is useful to know just which are the key life-history stages most affecting population growth. One can explore the sensitivity of the matrix model to each stage transition by changing each element one by one in a spreadsheet and observing changes in λ. Alternatively, elasticity matrices can be calculated whose elements represent the relative contribution of each element in the transition matrix to population growth (see Caswell, 1989, p. 79 for simple calculation methods adaptable for spreadsheet arithmetic). In the *Andropogon* example, the two smallest grass stages have by far the greatest effect on growth of the whole population.

Matrix models, despite the complexities of stage or age structure, share the assumptions of the simple geometric growth model and produce the same unlimited growth. Transition probabilities are assumed constant and density independent (though techniques are available for incorporating stochastic variation and density dependence, Usher, 1972; Caswell, 1989;

Box 4.1 The Clanwilliam cedar – a case study of the use of transition matrices for population modelling

Matrix modelling of fire-prone populations is difficult because transition probabilities change during a burn. A study of a rare conifer, *Widdringtonia cedarbergensis*, illustrates the problems (Manders, 1987). The tree is endemic to a single mountain range in the western Cape of South Africa. It has a life span of several hundred years but is easily killed by fire and the species has suffered an alarming decline over the past century. Manders used a stage-transition matrix to model population growth of *Widdringtonia* (see table opposite). The species takes about 15 years to reach maturity after which fecundity increases with size. The cedar does not have a persistent seedbank and some recruitment occurs between fires. The first departure from straightforward matrix models comes in predicting seedling establishment from seed production. Manders found that only 3 in 10000 seeds produced seedlings in unburnt stands, compared with 26 per 10000 after a burn, nearly an order of magnitude higher. Thus transition probabilities from seeds to seedlings have to change after fire. The second problem is predicting stage-specific mortality in a year with fire. This is obviously higher than the usual 'background' mortality. Each element in the column vector of size classes was multiplied by a corresponding element in a fire survival vector to simulate the effects of burning. However, survival differs among fires. Manders used a quadratic function relating survival to the length of the fire interval to provide 'a crude estimate of increasing mortality in fire with increasing vegetation age'. Thus transition probabilities for both seedling input and mortality change after fire.

The magnitude of fire-induced changes and their predictability are of critical importance in applying matrix models to fire-prone populations. In the cedar case, Manders predicted increasing populations if fires occurred every 20 years. However, intense fires have caused much higher mortality than used in the model – more than 80% of the trees in a 1989 summer wildfire. Fire intensity is not a simple function of vegetation age but depends on weather conditions and season of burn – event-dependent effects. Consequently, the predictive power of the demographic model is limited. At best it can be used to generate size distributions which can be matched against the predicted behaviour of a particular fire. Similar problems have beset attempts to model North American conifer forest dynamics (Kercher and Axelrod, 1984). No matter how elegant the mathematical modelling, fire, by its very nature, creates havoc with prediction.

(continued)

Table 4.1 Transition matrix model for *Widdringtonia cedarbergensis*, a long-lived conifer (ex Manders, 1987). The first row represents the mean number of seeds produced by a plant in that size class. The remaining elements give the probability of changing from one stage to the next. For example for class 2 seedlings, 74% will remain as seedlings over a period of one year, 18% will grow in height to class 3 (25–50 cm) and 2% will reach class 4 (50–75 cm). It can be inferred that 6% of the seedlings will die (i.e. 1 – (0.74 + 0.18 + 0.02))

Definition	*Class*	*Stages in tree growth*										
		1	*2*	*3*	*4*	*5*	*6*	*7*	*8*	*9*	*10*	*11*
Seed	1						2	34	110	510	1200	2800
Seedlings <25 cm	2	0.002	0.74									
25–50 cm	3		0.18	0.68								
50–75 cm	4		0.02	0.16	0.69							
75–100 cm	5			0.02	0.16	0.75						
100–150 cm	6			0.02	0.05	0.06	0.51					
>150 cm <5 cm DBH	7				0.03	0.04	0.37	0.83				
5–10 cm DBH	8							0.01	0.93			
10–20 cm DBH	9							0.003	0.04	0.93		
20–40 cm DBH	10									0.04	0.95	
>40 cm DBH	11										0.03	0.997

DBH, diameter at breast height.

Manly, 1990). The elegance of these models is seductive. However, introducing fire almost literally throws a spanner in the works, wreaking terrible havoc to a fine piece of mathematical machinery. All the transition elements have to be changed reflecting fire survival of the different stages and changes in post-fire reproduction (Box 4.1). Silva *et al.* (1991) got round the problem by defining two transition matrices – one for a burnt population and another for an unburnt one. They calculated λ to be less than 1 for the unburnt transition matrix, predicting a rapidly shrinking population in the absence of fire. *Andropogon semiberbis* apparently depends on frequent fires to persist.

A second example illustrating the difficulties of matrix models in fire-prone populations is shown in Box 4.1. The model was devised as a management aid for an endangered conifer with a life history peculiarly ill-suited to its fire-prone environment.

4.4 DEMOGRAPHY AND THE FIRE-INTERVAL HYPOTHESIS

There are a growing number of demographic studies on the effects of fire on plant populations. These use a variety of methods (of varying feasibility and reliability) for producing age- or stage-specific schedules (life tables) of death rates and fecundity. Each has advantages and disadvantages. The cohort approach traces survival and seed production of a single cohort through time. This is a convenient method for analysing decay rate of seedbanks but it is rarely used for perennial plants because they are so long-lived. It is often more convenient to measure survival, growth and fecundity of a group of plants for one to several years as in Silva *et al.*'s (1991) *Andropogon*, and the conifer example of Manders (1987), producing a 'time-specific' life table. The assumption here is that growing conditions over the measurement period are representative. Occasionally survivorship can be estimated from records of age at death, a common method in animal studies (Deevey, 1947). Schlesinger and Gill (1978) used an ingenious application of this method in studies of *Ceanothus megacarpus*, a fire-recruiting, non-sprouting chaparral shrub which is therefore even-aged. Standing dead stems of *Ceanothus* persist in chaparral stands for decades. From counts of year rings in these dead stems, survivorship curves could be calculated by estimating the number of plants surviving after each year since fire. Unfortunately the long persistence of dead stems is a prerequisite for this method and limits its wider application.

One of the most common methods used in fire studies is the 'chronosequence' – the comparison of populations across stands of different post-burn ages (Gill and McMahon, 1986). In long-lived plants chronosequences often seem the most practical tool for population analysis. However, their limitations should be clearly understood and controlled wherever possible. The assumption is that population age is the only significant variable differing among the stands, often violated in patchy habitats or stands of unknown fire history. Chronosequences across different soil types or different slopes and aspects will confound age and site differences and should be avoided. Differences in stand density are also often overlooked and may confound age and density effects.

We first consider how survival patterns vary among plant populations and why they are affected by the length of the fire interval. We then turn to reproduction and the ecology of seedbanks and their dependence on time since last fire. We treat the demography of woody plants separately from grasses since these two growth forms behave very differently.

4.4.1 Survivorship patterns

One of the major differences between plant and animal demography is longevity. Individual plants, in some species, can persist for literally thousands of years. This capacity for individual persistence can confer great stability on populations and inertia in the communities containing them. Fire often kills plants or plant parts, creating gaps and the opportunity for vegetation change. An understanding of how plant persistence changes under different fire regimes is therefore of central importance to understanding vegetation dynamics.

Non-sprouters are the least persistent perennials and therefore most likely to have unstable populations. Much of this instability relates to the state of the seedbank when the population is burnt. Sprouters show great variability in individual persistence. Some species are extremely persistent, effectively immortal in the case of clonal root-suckering species. Others are nearly as ephemeral as non-sprouters, with high fire-induced mortality. Since survival response is often age- or size-dependent, populations may be sensitive to variation in the fire interval.

We first discuss the importance of fire interval in non-sprouter populations and then its effect on sprouters.

Fire-recruiting non-sprouters

Fire-recruiting non-sprouters have single-aged populations dating from the last fire. If the interval is too short, populations may be killed before they have acquired sufficient seed to replace themselves. If the interval is too long, populations may have thinned to such an extent that they also cannot replace themselves. Where dormant seedbanks persist after the death of parent plants, the fraction surviving would be expected to decline until the next burn.

There are marked differences among species with respect to life spans and survivorship curves. In many systems short-lived species are abundant for the first few years after fire then die and persist only as soil-stored seedbanks. This life history is particularly common in legumes, including *Acacia* species in Australia (Auld, 1986), *Lotus scoparius* in chaparral (Keeley, 1991), *Aspalathus* spp. in fynbos (Kruger, 1977) and a variety of shrubby and herbaceous legumes in Mediterranean shrublands and woodlands (Naveh, 1975; Papavassilou and Arianoutsou, 1993).

Trees and shrubs with longer life spans have a tighter dependence on population status at the time of fire. Survivorship curves for a number of non-sprouting shrubs are shown in Figure 4.6. The heaviest mortality occurs in the first season of growth. Subsequent survivorship and longevity varies greatly among species and between populations of the same species. In *Protea neriifolia*, a serotinous fynbos shrub, mortality was concentrated towards the end of the life span. *Orothamnus zeyheri*, a fynbos shrub with soil-stored seeds, and *Ceanothus crassifolius*, a chaparral shrub, had relatively constant mortality rates over the study period. *Banksia ornata*, *B. ericifolia* and *C. megacarpus* had high early mortality rates but low subsequent rates. These interspecific differences in survivorship could produce

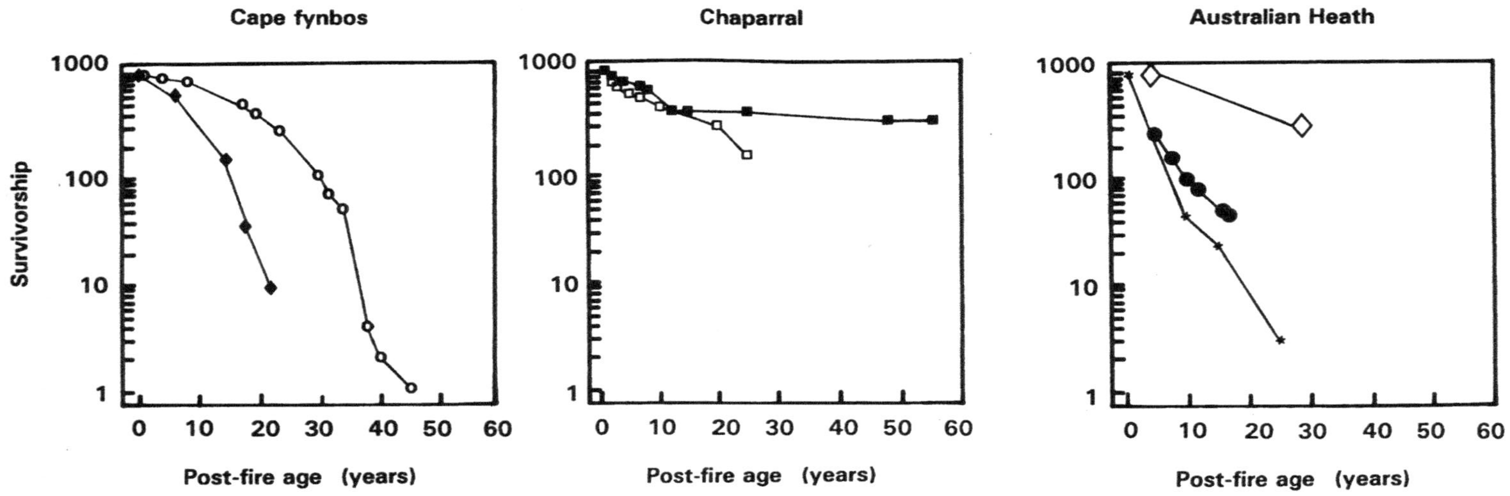

Figure 4.6 Survivorship curves for a number of fire-recruiting non-sprouting shrubs in Cape fynbos, Australian heath and Californian chaparral (from Le Maitre and Midgley 1992). Cape fynbos: ◆, *Orothamnus zeyheri*; ○, *Protea neriifolia*. Chaparral: ■, *Ceanothus megacarpus*; □, *C. crassifolius*. Australian heath: ◇, *Banksia ericifolia*; ●, *Acacia suaveolens*; *, *Banksia ornata*. *Acacia suaveolens*, the two *Ceanothus* species and *Orothamnus zeyheri* all have soil-stored seedbanks. The remaining species are serotinous.

different responses to fire frequency – unless compensated by differences in seed production or seedbanks.

Senescence

Fire-recruiting non-sprouters often senesce and die in the long absence of fire, producing a mass dieback. This phenomenon was first described by Watt (1947) for *Calluna* heathlands in Britain. He recognized four phases in the life of a *Calluna* plant: pioneer (up to 6 years), building (up to 15 years), mature (up to 20 years) and degenerate with central branches dying and creating a gap. Death could follow at 30–40 years or, as Watt believed, a new growth cycle could begin by seedlings establishing in the gap or by vegetative spread and axillary rooting of neighbouring shrubs. The cyclical nature of *Calluna*'s growth is not universal and the degenerate phases may be colonized by trees or bracken, so replacing the heath (Gimingham, 1988).

This pattern of senescence is found in many woody species experiencing regular crown fires in different parts of the world. In Cape fynbos, for example, the dominant proteoid shrubs collapse and die after 30–60 years in the absence of fire. In serotinous species, the seedbank does not persist after the death of the parent, and local extinction can result from the senescence of a stand (Bond, 1980). Sometimes, as in *Calluna*, seedlings establish successfully in the gaps created by dead parents, though in much reduced numbers. Similar patterns of senescence have been reported for *Ceanothus* species in California (Horton and Kraebel, 1955), shrubs in Australia (Cowling *et al.*, 1990) and serotinous conifers (*Pinus contorta*, *P. resinosa*) in North America.

The causes of senescence are poorly understood. Keeley (1986) has argued, for chaparral shrubs, that individuals senesce but populations do not. He cited the case of *Ceanothus megacarpus* where high initial mortality appeared to be due to density-dependent thinning, but subsequent mortality rates were very low, even in very old stands. Life tables typically ignore density effects on mortality rates and Keeley's argument deserves close scrutiny. Kruger (1986), however, found no evidence for density-dependent mortality in a fynbos *Protea* and this species showed typical programmed senescence.

There is considerable interest in the evolutionary causes of senescence and a diversity of possible explanations (Chapter 6). Loehle (1988) has collected data from many trees indicating a trade-off, in the form of a positive correlation, between age at first reproduction and longevity. He attributed the correlation to the competing demands on plant resources of investment in reproduction or in decay-resistant wood. There has been some attempt to test for such a trade-off in chaparral (Keeley and Zedler, 1978) and Cape fynbos shrubs (Frost, 1984) but the data are weak. It would be interesting to test the relationship between longevity and reproduction in fire-prone environments (Chapter 6).

Sprouters

Of key importance in the demography of sprouting species is size- or age-dependent variation in the capacity to sprout. The length of the fire interval affects population growth because sprouting is size-dependent. Typically, sprouting is weakest in the smallest size classes. Short fire frequencies may therefore eliminate new recruits before they have acquired sprouting capabilities. Fire interval also affects architectural changes. Many tree species, especially in savannas, pass through a multi-stemmed stage before becoming single-stemmed (Pellew, 1983; Menaut *et al.*, 1990). Short return intervals maintain plants in the stunted, multi-stemmed form where they are subject to browsing and competition from grasses (Chapter 7).

Stage-specific fire survivorship curves are quite variable among species. It is convenient to divide this variability into four major patterns (Figure 4.7). The seedlings of all species have a high probability of being killed. Some species, such as clonal plants with the capacity to sprout from roots, very rapidly acquire resistance to burning and this is maintained across all size classes (Figure 4.7, type I). Others become increasingly resistant to fire as they grow larger but are vulnerable over a larger range of sizes (Figure 4.7, type II). Many species rapidly acquire fire resistance as juveniles but become more prone to fire damage in the largest size classes (Figure 4.7, type III). For example, *Baccharis pilularis* ssp. *consanguinea*, a Californian shrub, lost its capacity to resprout after 7 or 8 years apparently because regeneration buds in juvenile tissue were overgrown by secondary stem growth (Hobbs and Mooney, 1985). Similarly chamise, *Adenostoma fasciculatum*, mortality was positively correlated with size after a burn but negatively correlated with size between burns (Figure 4.8). Savanna trees often form fire-resistant multi-stemmed shrubs as juveniles but lose the capacity to sprout when they are large, single-stemmed trees. Finally, type IV species, non-sprouters, are always killed by fire, regardless of size class.

These size-dependent mortality patterns implicate fire interval as a key determinant of population change. However, the level of mortality, especially in type II and type III fire survivorship patterns, can vary greatly from fire to fire regardless of fire interval. These event-dependent effects are discussed in Chapter 5. There seem to be rather few studies of variation in size-dependent sprouting among co-occurring species though it could be informative for predicting vegetation change. For example, Hodgkinson (1986) showed that contrasting size-dependent mortality patterns in three shrubby species from a semi-arid *Eucalyptus* woodland were good predictors of cumulative mortality in successive annual spring burns. Population size of a species with type IV mortality declined the most rapidly, followed by a species with type II mortality. The least change occurred in a species with a type I pattern of mortality.

In some South African savannas, *Acacia* species follow type III curves,

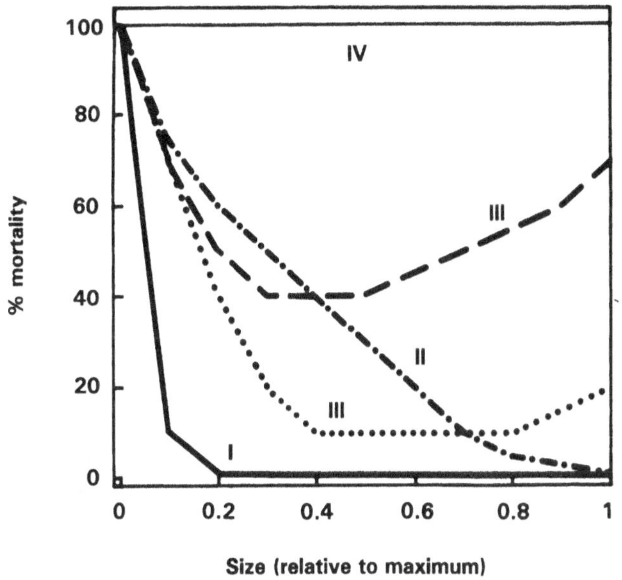

Figure 4.7 Fire-induced mortality patterns in relation to plant size. Four broad patterns can be recognized: I, very low post-seedling mortality; II, decreasing mortality with increasing size; III, mortality highest in smallest and largest size classes; and IV, non-sprouters killed across all size classes. The level of mortality can change greatly from fire to fire, especially in type III plants, shown here by two curves representing high- (upper curve) and low-intensity (lower curve) burns.

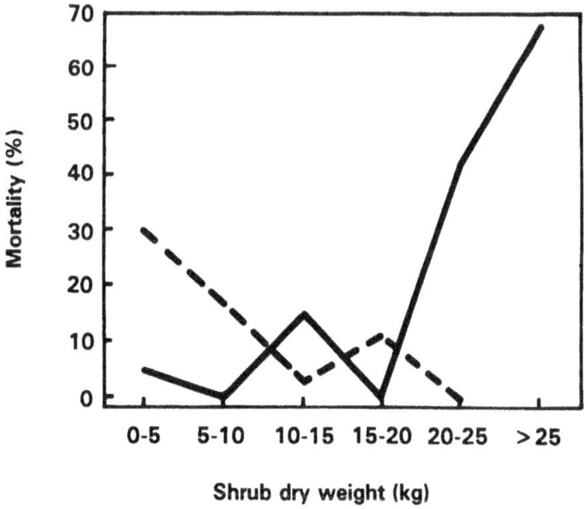

Figure 4.8 Size-dependent mortality patterns in a resprouter, *Adenostoma fasciculatum*. Small size classes have the heaviest mortality between fires (dashed line) while large size classes are more likely to be killed by a fire (solid line). (From Stohlgren and Rundel, 1986.)

becoming increasingly vulnerable as they age. With long periods without fire, *Acacia* stands are invaded by broad-leaved shrubs (e.g. *Euclea)* with type I responses which rapidly acquire extreme resistance to burning (Smith and Goodman, 1987). Once *Euclea* has invaded an *Acacia* stand and grown to a resistant stage, it forms an extremely persistent component of the community which cannot easily be eradicated.

Savanna trees

Demographic studies of grass and savanna trees have been neglected until recently. Savanna trees and grasses are intimately connected through fire and the effects of herbivores (Chapter 7). Many savanna trees resprout as juveniles or revert to resprouting from the base if the canopy is burnt (Trollope, 1984). In this stunted state they compete with grasses and are more heavily browsed. They can be prevented from emerging from the grass layer by regular burning. If fires are infrequent or of low intensity, trees grow to the point where they overtop grasses, reducing the risk of future fire. Stage-specific growth rates and the probability of reversion to previous stages due to fire or herbivory are thus key demographic parameters for predicting tree response to variable fire frequency and intensity (Dublin *et al.*, 1990; Menaut *et al.*, 1990; Chapter 7). The system would seem readily amenable to transition matrix modelling.

Modelling sprouter demography

Demographic analysis of plants that survive fires by sprouting is complex. Life-table analyses of survival, growth and reproduction of both plants and seedbanks must be calculated for the interval between fires. Demographic studies of sprouting plants have been plagued by the difficulty of determining age. There have been several ^{14}C isotopic estimates of lignotuber age in woody plants (Table 4.2) but the complex anatomy of lignotubers makes accurate age estimates difficult. Studies of resprouters have therefore concentrated either on the age structure of resprouting stems (ramets) or stage-specific survival of fire.

Stohlgren and Rundel (1986) used stage-specific data on growth and fire survival to model the demography of *Adenostoma fasciculatum*, a resprouting chaparral shrub (see opposite). *Adenostoma* had a very similar size structure in different chaparral stands in their study area, suggesting very stable populations from one fire to the next (Figure 4.9). By assuming a stable size structure, stable population size and a surplus of seedlings over establishment sites, Stohlgren and Rundel were able to derive a survivorship curve for this species from which they inferred that some shrubs can survive more than 23 fires, representing a maximum life span of between 720 and 1920 years.

In some species individuals are so persistent that attention has focused on the demography of ramets, rather than genets. Some resprouting species

Table 4.2 Longevity of lignotubers, radiocarbon dates

Eucalyptus oleosa	Australia	190 (±60) yr 330 (±70) yr	Wellington *et al.* (1979)
Adenostoma sparsifolium	Chaparral	200 (±65) yr	Hanes (1965)
Audouinia capitata	Fynbos	90 (±10) yr	De Lange and Boucher (1992)

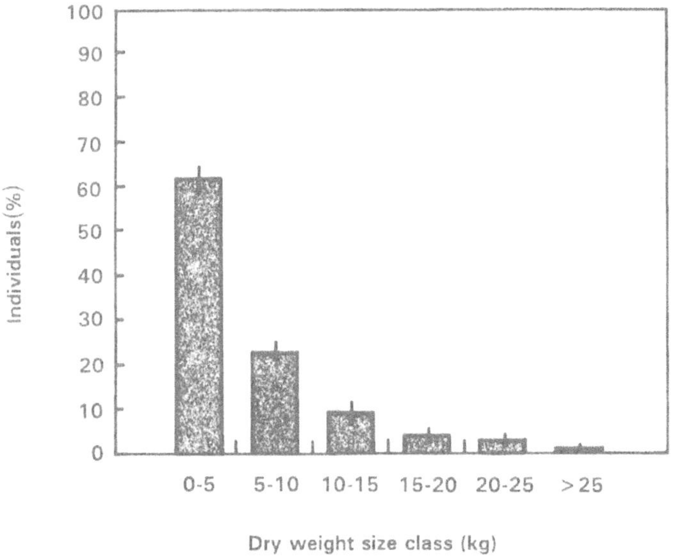

Figure 4.9 Stable size structure of *Adenostoma fasciculatum*, a chaparral resprouter. Note the very low standard errors in size classes recorded in 10 separate sites, suggesting a stable size class distribution and little variation in population size from one fire to the next (Stohlgren and Rundel, 1986).

continually rejuvenate their canopy by initiating new sprouts from their bases in the absence of fire. This implies great resilience to long fire-free intervals. Keeley (1986) noted that non-fire-recruiting sprouters ('obligate sprouters') required very long fire-free intervals for successful seedling recruitment in chaparral. This was in marked contrast to fire-recruiting sprouters ('facultative sprouters') such as *Adenostoma* which benefit from frequent fires. The implications for species coexistence is discussed further in Chapter 7.

Grasses

Grasses are also affected by fire interval but not for the same reasons as in woody plants. Annual dieback of grass leaves can produce a dense litter

Table 4.3 The effects of shading on tiller initiation in six grass species from the humid grasslands of Natal. Values are the increase of tillers relative to plants grown in pots in full sunlight. Decreaser species decline in abundance with long intervals between fires in contrast to increaser species. (From Everson *et al.*, 1988)

Shade treatment (% full sunlight)	*Decrease with long fire intervals*			*Increase with long fire intervals*		
	Themeda triandra	*Heteropogon contortus*	*Trachypogon spicatus*	*Harpechloa falx*	*Tristachya hispida*	*Alloteropsis semialata*
80	56.4	36.8	26.5	24.7	26.7	18.9
60	12.6	4.5	4.6	17.5	19.3	6.9
30	1.4	0.0	0.0	7.7	1.7	0.5

layer if the surplus is not removed by fire or grazing. Different grass species vary in their tolerance of this litter layer. Some bunch grasses become moribund without fire because of self-shading of tillers formed at the basal nodes. Others are relatively unaffected by dense shading (Table 4.3; Everson *et al.*, 1988).

In the humid grasslands of Natal, shade-tolerant species have swollen bases and initiate tillers below ground, whereas shade-intolerant species produce tillers from nodes elevated above the ground and lack swollen bases (Everson *et al.*, 1988). These species differences are important for determining the effects of fire interval on species composition – long fire intervals favour shade-tolerant species at the expense of shade-intolerant ones (Tainton and Mentis, 1984).

As discussed earlier in this chapter, grass populations can be conveniently modelled using transition matrices. By measuring the fate of tillers, Silva *et al.* (1991) used a matrix modelling approach to show that a South American savanna grass was very sensitive to the length of the fire interval. Without regular fires, populations of this species would rapidly decline. Using a similar approach, Everson *et al.* (1985) developed transition matrices for an African grass burnt under three different fire regimes. Here, however, fire interval had less of an effect than fire season. Annual winter burns produced the highest predicted population growth rates. Biennial spring burns produced lower growth rates, while annual summer burns were predicted to result in a decline in the population. These modelling results, based on short-term (2 year) demographic studies of the fate of tillers, were consistent with patterns of relative abundance observed after a 30 year burning experiment.

These matrix models assume exponential growth, and do not take into account factors such as self-shading by litter. The inclusion of density-dependent feedbacks in grass population models has interesting consequences for population stability which are discussed further in section 4.5.4.

4.4.2 Fecundity

Age-specific fecundity is of central importance in the demography of non-sprouting species. Fires that occur before a population has reached maturity can cause local extinction. Fires that occur after a population, or its seedbank, has senesced will have the same effect. For non-sprouting species, minimum juvenile period and maximum life span are key demographic parameters for determining the effects of fire interval on population persistence. However, it is less certain whether fire interval significantly affects fecundity, or recruitment, between these critical periods.

There are comparatively few studies of age- or size-specific fecundity in woody plants because of the vast quantity of seeds produced by most long-lived trees. However, in the nutrient-poor heathlands of South Africa and Australia, seed production can be remarkably low, especially in large-seeded Proteaceae. An example of a comparative study of four *Banksia* species is illustrated in Figure 4.10. These species had a pre-reproductive juvenile period of 4–6 years followed by an exponential increase in flower number declining eventually as plants approached their full life span. The length of the juvenile period is of critical importance since fires that occur before reproductive maturity cause local extinction.

Once a population has reached maturity, age or stage is not necessarily a good predictor of fecundity. Other factors, especially rainfall, also influence annual reproduction. Keeley (1977a) has studied seed production in congeneric pairs of sprouting and non-sprouting *Ceanothus* and *Arctostaphylos*. Over a 3 year period, *Ceanothus* fruit production varied from less than 10 per m^2 of plant cover to 2454 in *C. leucodermis* and 5252 in *C. greggii*. Over a 5 year period *Arctostaphylos* fruit production varied over three orders of magnitude. Keeley found a striking correlation between fruit production and the previous year's rainfall for both *Arctostaphylos* species, and suggested that this relationship is widespread in chaparral shrubs. Pierce and Cowling (1991) found three- to tenfold differences in seed production of six fynbos shrubs in two successive years and a seventh species with relatively constant production. Such variable seed production suggests considerable caution in predicting fecundity from age or stage alone. It implies that fire frequency may not always be a good predictor of seed availability once the population has reached maturity.

Seedbank dynamics

The demography of plants cannot be understood without consideration of the fate of seeds (Figure 4.2). Survivorship curves for seeds are at least as important as fecundity and survival rates of vegetative plants for understanding population dynamics. Short-lived fire ephemerals spend most of the fire cycle as seeds in dormant seedbanks. Depending on seed decay

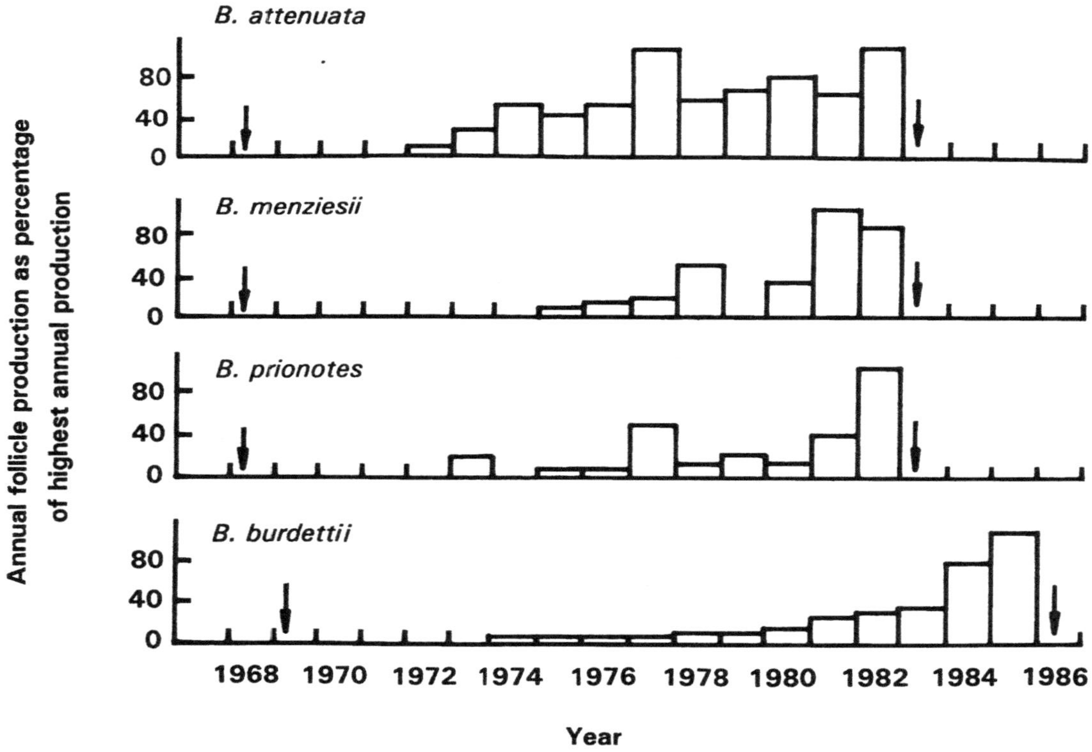

Figure 4.10 Age-dependent fecundity (follicle production) in four *Banksia* species in the northern kwongan of south-western Australia. *Banksia attenuata* and *B. menziesii* are sprouters and *B. prionotes* and *B. burdettii* are non-sprouters. The arrows indicate fire dates. (From Cowling *et al.*, 1990.)

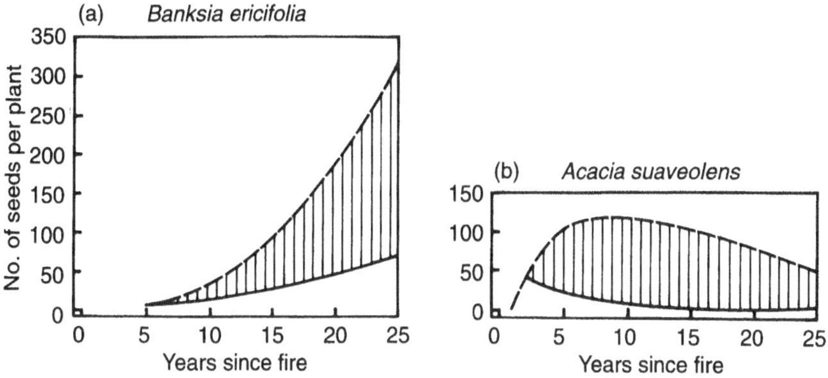

Figure 4.11 The relative contribution of current seed production (solid lines) and stored seeds (dashed lines) to total seed reserves. The shaded area indicates the contribution of the seedbank to total seed reserves. (a) *Banksia ericifolia*, with canopy-stored seeds (from Bradstock and O'Connell, 1988); (b) *Acacia suaveolens* with soil-stored seeds (from Auld, 1986).

rates, the length of the fire interval could determine the number of seeds surviving until the next fire.

The ecology of seedbanks is notoriously difficult to study. Serotinous plants are a conspicuous exception since the seedbank is visible on the plant. Bradstock and O'Connell (1988) have described seedbank dynamics in *Banksia ericifolia*, an obligate non-sprouter, using a combination of chronosequence and cohort analysis. Plants first flowered after 5 years. Mean seed production per plant was estimated as 4.408 (1.116^t) where t is the number of years since the previous fire. Seed survival was estimated as $S_{tj} = 0.8663^t$. Combining the two (Figure 4.11a) shows that seedbanks contributed substantially to total seed reserves with the relative contribution increasing as the stand aged. Thus serotinous follicles stored a little over twice the current seed crop in 10-year-old plants and more than three times the current crop in 20-year-old plants.

There is considerable variation in seed loss rates and life spans in seedbanks, as illustrated by the growing number of studies of serotinous Proteaceae. The significance of this variation is poorly understood. This is an area of fire ecology where the data far exceeds interpretive tools. Clearly seedbanks must be studied together with life tables of plants to provide a complete picture.

Soil seedbanks

The diversity of fire life-histories shown in Figure 4.2 implies diverse dynamics of soil-stored seeds. Seedbanks of fire ephemeral species should decline over long fire intervals since there will be no fresh input. Auld (1987) studied soil seedbanks formed by an early successional

Australian shrub, *Acacia suaveolens*. He found an exponential decline in seedbank size once seed production ceased (Figure 4.11b). This species would benefit from short fire frequencies. Nevertheless, some viable seeds persist in the soil for decades and some seedlings would emerge even after long intervals without fire. Zammitt and Zedler (1988) studied seedbanks of a variety of species in Californian chaparral. Herbs could be broadly divided into species whose seedbanks declined over time and an opposing group whose seedbanks increased as stands aged. The former were post-fire ephemerals with a single seed input in the first year after fire. The latter were annuals which do not require fire to germinate and can grow and reproduce in mature chaparral, continually adding seeds to the seedbank.

Soil seedbank dynamics of long-lived plants appear very diverse but we do not know why. Keeley (1977a) found a relatively constant soil seedbank in *Ceanothus* and *Arctostaphylos* species despite orders of magnitude variation in annual seed supply. The annual surplus is presumably lost to predators, both vertebrate and invertebrate. Zammitt and Zedler (1988) also found no relationship between chaparral age and seedbank size in *Ceanothus greggii*. Pierce and Cowling (1991) studied soil seedbanks in a number of shrubby fynbos species. As in the chaparral examples, persistence times were comparatively short so that the species were apparently dependent on the most recent years' seed inputs for post-fire recruitment. Fire frequency would be a poor predictor of changes in abundance in these species. The pattern is quite different in *Adenostoma fasciculatum*, a resprouter. Zammitt and Zedler (1988) found increases from 16 to 51 to 150 seeds dm^{-2} in 9-, 35- and 85-year-old stands, respectively. This pattern of increasing seed densities over time has also been reported for *Calluna vulgaris* in British heathlands (Mallik *et al.*, 1984). Neither species has pure fire-stimulated seed germination and interfire recruitment can occur. These contrasting patterns of seed accumulation have interesting, but so far untested, implications for population regulation by fire frequency.

Grass fecundity and seedbanks

Many grasses flower profusely after fire (Daubenmire, 1968) with declining reproduction in subsequent years. The seed biology of grasses in relation to fire is poorly known. Some species have hygroscopic awns which result in self-burial and presumably increased fire survival. However, the relative importance of fire-stimulated flowering, seedbank dynamics and seedling emergence for population growth are not well studied.

4.4.3 The disturbance-interval hypothesis and population regulation

How important are varying fire intervals for regulating plant populations? The evidence is very uneven. Fire interval determines the likelihood that

Table 4.4 Key stages in fire life-histories. Population growth is strongly affected by the length of the fire interval in relation to these stages (Noble and Slatyer, 1980; Chapter 7). Survival through a stage also depends on the particular characteristics of a fire (Chapter 5). Stages marked with an asterisk are not universal to sprouting species but depend on stage-specific fire survival patterns

Life-history type	*Key life-history stage*	*Effect of fire interval*
Non-sprouter	Age at first reproduction	No seeds at short intervals
	Maximum life span	No seed input over intervals longer than the life span
	Maximum seed life span	No seeds after intervals longer than the life span + seed life span
Sprouter	Stage at which capacity to sprout is acquired	No recruitment at shorter intervals
	Stage of transition from multi- to single-stemmed*	Escape from understorey suppression and crown damage at longer intervals
	Stage when capacity to sprout is lost*	Fire-induced mortality of large individuals

plants will pass key hurdles in their life history (Table 4.4). For non-sprouters the hurdles are age of first reproduction, maximum life span and longevity of seeds. Qualitative predictions on the fate of a species under different fire frequencies can be based on these key parameters alone (Table 4.4; Noble and Slatyer, 1980; Chapter 7). However, quantitative evidence for the importance of fire interval in the period between these key times is scarce.

Demographic studies, usually designed to aid fire management, have been essentially descriptive. Predictions arising from such studies have very seldom been tested. Seedbank studies, especially of serotinous species, show age-dependent changes in seedbank size (e.g. Figure 4.11). Population growth in these species should therefore depend on fire frequency. However, other factors intervene. Thus Bradstock and O'Connell (1988) predicted a tenfold increase in recruitment of *Banksia ericifolia* and *Petrophile pulchella* if fire intervals were increased from 6 to 15 years, but a 30-fold increase by switching fire season and post-fire weather (Figure 4.12).

Lamont *et al.* (1991b), in a study of a Western Australian *Banksia*, found that the weather pattern following a burn best accounted for recruitment patterns, rather than size and dynamics of the canopy seedbank.

For sprouters, key life-history hurdles depend on size-specific fire survival patterns (Table 4.4). All sprouters can be killed by fire when they are small. Successful recruitment thus depends on sufficient time to pass the

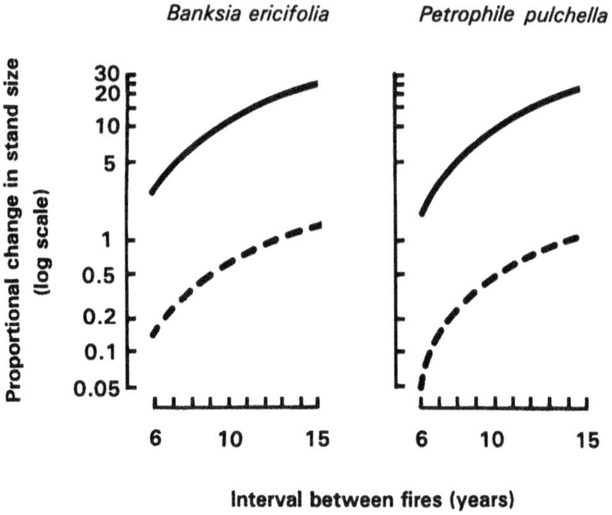

Figure 4.12 Predicted population growth as a function of fire interval and 'good' (solid line) v. 'bad' (dashed line) fires for *Banksia ericifolia* and *Petrophile pulchella* from eastern Australia. The large differences in the two fire types are due to different fire seasons and post-burn moisture status (Bradstock and O'Connell, 1988).

minimum sprouting threshold. Mesquite (*Prosopis*) seedlings sprout vigorously after 18–24 months (Wright *et al.*, 1976). Recruitment can therefore be prevented by annual burns but not by longer fire frequencies. Some sprouters lose the capacity to sprout when they are large. The stage at which this happens is a second important hurdle influencing fire-interval effects. However, size-specific survival patterns are also dependent on fire intensity, fire season and other event-dependent effects, such as the destruction of bark by animal activity (Chapter 5). In grasslands, fire season is often as, or more, important than fire interval in driving population change (Chapter 5).

The disturbance interval hypothesis is implicit in many fire studies. Qualitative predictions based on key life-history stages are an important tool for predicting vegetation response to changing fire frequencies (Table 4.4). The available demographic studies support the importance of the timing of key life-history events. However, much of the variation in population size cannot be attributed to fire frequency and other factors deserve closer attention.

4.5 DENSITY DEPENDENCE AND POPULATION REGULATION – THE SELF-REGULATORY HYPOTHESIS

Life-table studies, whether of populations or seedbanks, often assume that growth, survival and fecundity vary only with time and are independent of population density. We have already seen that the geometric growth equations with constant λ generate ridiculous numbers without some form of regulation. Density-dependent variation in life-history parameters is one of the most common forms of population regulation. Is there evidence for density-dependent population regulation in fire environments, and what form does it take?

4.5.1 Mortality

Fire-recruiting non-sprouters are well suited to studies of intraspecific mortality. Initial seedling numbers can vary widely after a fire but subsequent mortality can have a density-dependent component. There are several approaches to demonstrating density-dependent mortality. As a population ages, the initially variable densities should start to converge as a result of self-thinning. The way in which self-thinning occurs has been labelled the –3/2 self-thinning rule. This states that the relationship between size and density is such that a double logarithmic plot of mean weight per plant against stand density is linear with a slope of about –1.50 – hence the –3/2 law (Yoda *et al.*, 1963). The law predicts that plants which are too small or at low densities grow without thinning until they reach the –3/2 slope, when size increases at the expense of density. There is an extensive literature on the rule but growing scepticism as to its generality (Weller, 1987).

Another way of viewing density dependence is to study patterns of interactions between individual members of a population rather than stand averages (Pacala and Silander, 1985). Such studies lead to several predictions if density-dependent competition is occurring. First, in an even-aged stand such as occurs after fire, variation in size might be reduced with time as a result of asymmetric competition for light. Size distributions would be expected to change from highly skewed with a few dominants and many suppressed individuals to an approximately normal distribution as the suppressed individuals die (Weiner, 1990). Secondly, if the probability of death of a plant is a function of local density, then distribution patterns should also change with time from random or clumped to more regular as thinning removes suppressed competitors. The most crowded plants should have the highest probability of dying.

Schlesinger and Gill (1978) have studied most of these aspects of density-dependent thinning in *Ceanothus megacarpus*, a chaparral shrub. In their intensive 21-year-old study site, self-thinning halved densities from 6 to 3 stems m^{-2} over a 16 year period. They found a slope of –1.23 for a double log plot of mean plant weight against stand density, rather than the –3/2 power law of Yoda *et al.* (1963). At lower densities, the distribution of plant sizes changed from highly skewed towards normal, and thinned plants were drawn from smaller size classes as predicted (Figure 4.13). The variance to mean ratio, a measure of pattern, indicated a shift from contagious towards random and regular distributions. Thus *C. megacarpus* provides evidence

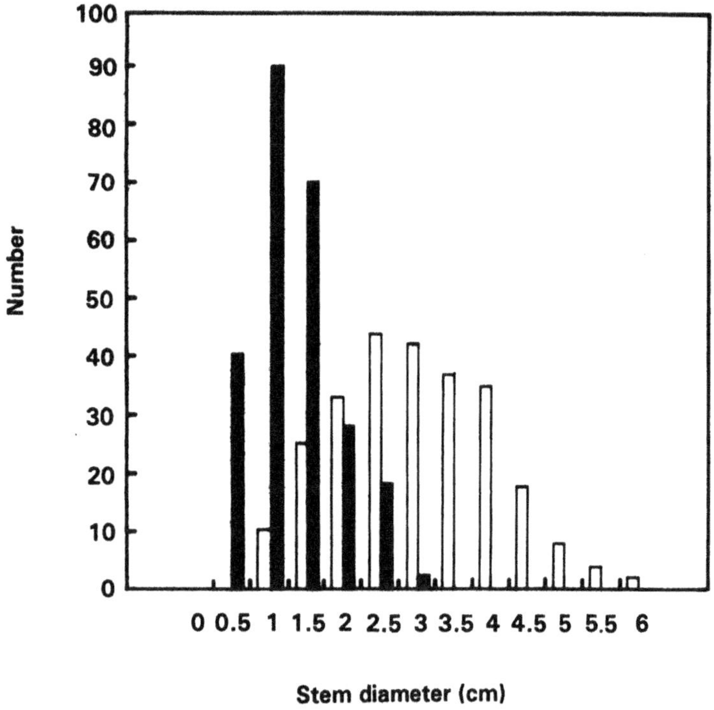

Figure 4.13 Size class distribution of living (open bars) and dead ('thinned') plants (shaded bars) of *Ceanothus megacarpus*, a chaparral shrub (from Schlesinger and Gill, 1978).

for density-dependent mortality due to intraspecific competition. In a related species, *C. crassifolius*, dead shrub density was linearly correlated to stand density but this density-dependent mortality did not lead to convergence on an asymptotic density as predicted by the –3/2 law (Riggan *et al.*, 1988).

Recent studies have related individual tree deaths to the space available for each individual. Kenkel *et al.* (1989) studied jack pine (*Pinus banksiana*), a serotinous non-sprouter of boreal forests. Like *C. megacarpus*, populations of this species changed from random or clumped to regular as stands aged and plants thinned (Kenkel, 1988). Kenkel *et al.* (1989) mapped individual trees in a 65-year-old even-aged stand, classed as live, standing dead or stump. Each tree occupied a polygon (assumed to approximate its area of available resources) defined by lines bisecting distances to nearest neighbours (Figure 4.14). They found that polygon areas of surviving trees were significantly larger than those of dead trees and that trees with larger polygon areas had much higher probabilities of survival (Figure 4.15). These studies provide good evidence for density-dependent regulation by 'contest' competition for space.

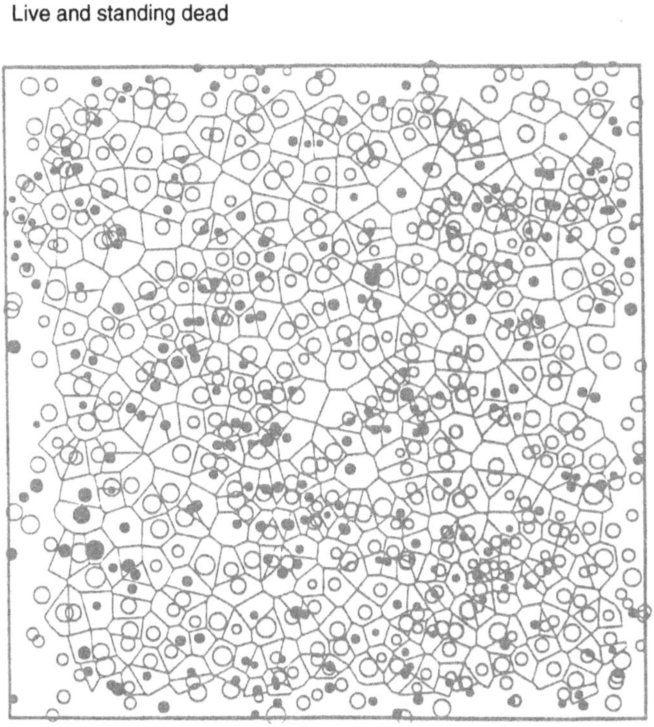

Figure 4.14 The influence of neighbourhood size on plant mortality in jack pine (*Pinus banksiana*). Lines bisecting the distance between all possible neighbours were joined to produce polygons defining neighbourhood size. The pattern map for live and standing dead trees is shown. Open circles represent surviving trees and solid circles, dead individuals. Circle size is proportional to diameter at breast height (DBH). (Reproduced from Kenkel *et al.*, National Research Council of Canada, 1989.)

4.5.2 Fecundity

Harper (1977) has reviewed evidence for density-dependent regulation of plant reproductive effort. In general, as density increases, the dimensions of an individual plant and the number of flowers, fruits and seeds decline. The population consequences depend on the product of individual reproductive effort and plant density. As population density increases, seed production per plant may:

- remain constant for the space available, resulting in constant seed yield for the population;
- increase for the space available, resulting in increasing seed density (called **under-compensation** since plants under-compensate for diminishing resources); or
- decrease with the space available so that seed yield for the population

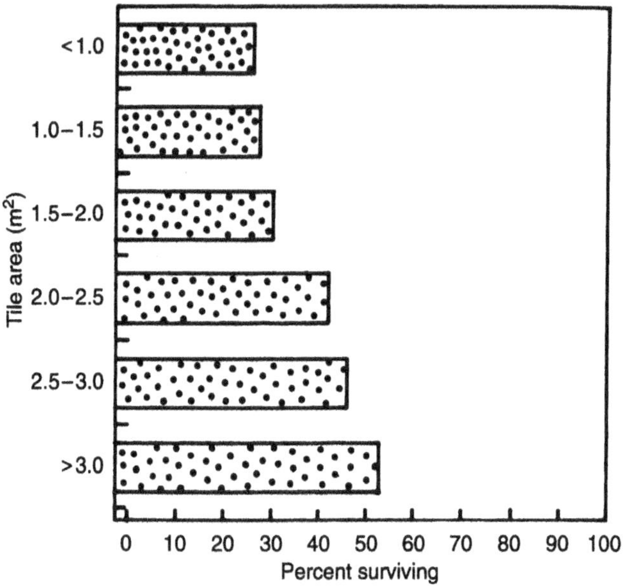

Figure 4.15 Survival as a function of neighbourhood size in *Pinus banksiana*. Tile area refers to the polygons described in Figure 4.14. (Reproduced from Kenkel *et al.*, National Research Council of Canada, 1989.)

declines as density increases (called **overcompensation** since plants overcompensate for diminishing space).

The different reproductive responses to shrinking space have important implications for population regulation (May, 1974; Bellows, 1981; Lomnicki, 1988). Exact compensation would tend to stabilize population growth, whereas overcompensation, with its potential for strong negative feedback, could result in chaotic dynamics.

Studies of density-dependent effects on reproduction have generally been neglected in studies of fire-prone species, although serotinous plants offer exceptional opportunities for measuring fecundity in long-lived species. Bond *et al.* (1984) found evidence for density-dependent regulation of reproduction in *Protea lorifolia*. As in other studies (Harper, 1977; Rees and Crawley, 1989), there was a log–log relationship between plant size and the number of inflorescences per plant but only once the shrubs had crossed a size threshold. The consequence was that inflorescence number declined as plants became crowded. Maze and Bond (1995) have explored density-dependent effects on cone production in two species of *Protea* (Figure 4.16). For both species there was an optimal density for cone production above and below which cone numbers declined. The decline in fecundity at high stand densities was attributed to the persistence of many individuals which competed for space but were sterile. It is easy to see that the shape of

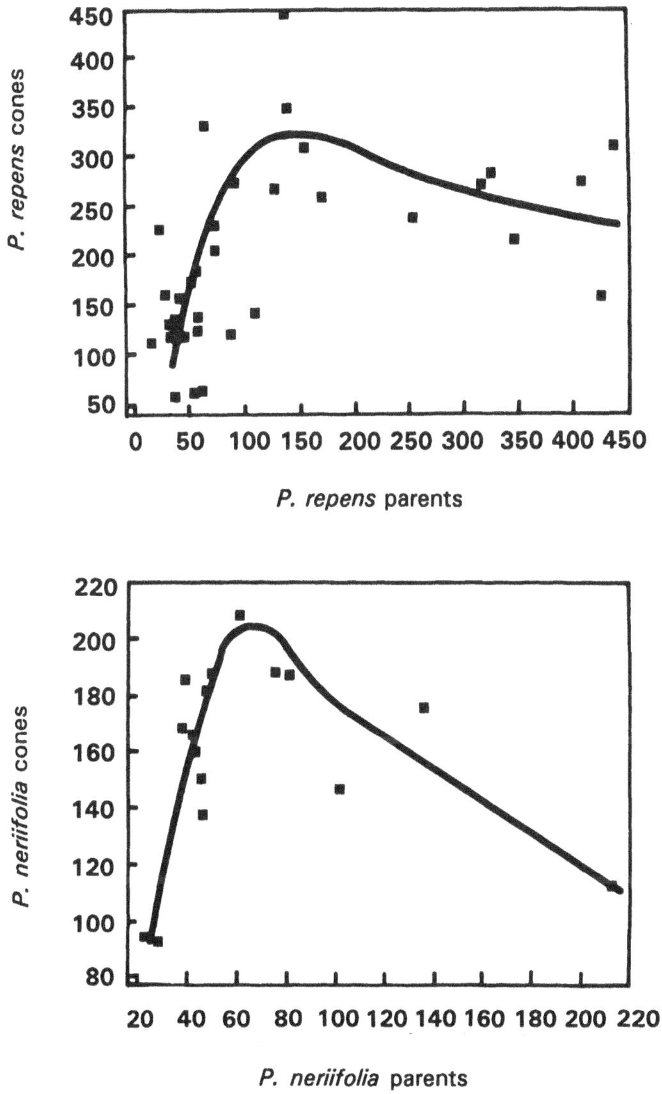

Figure 4.16 The relationship between cone production per plot (10 × 10m) and plant density for two species of fynbos proteas (from Maze and Bond, 1995).

the density/fecundity curves could have implications for the stability of these non-sprouting populations over successive fires (see below).

4.5.3 Effects of density dependence on population regulation

There is growing interest in the relationship between density-dependent regulatory mechanisms and plant population dynamics (Symonides *et al.*, 1986; Thrall *et al.*, 1989; Silvertown, 1991). Chaos and oscillatory dynamics as a result of strong density dependence may be rare in plant populations because of the stabilizing effect of self-thinning on mortality and the modular structure of plants on fecundity (Rees and Crawley, 1989). Many populations in fire-prone systems seem relatively stable. Stohlgren and Rundel (1986), for example, found remarkably constant size class distributions of the sprouting shrub *Adenostoma fasciculatum* in mature Californian chaparral despite recurring disturbance by fire (Figure 4.9). This species produces abundant seedlings after fire. Interfire thinning of crowded seedlings and replacement of small shrubs killed by fire appears to maintain the stable population structure. Changes in fire frequency or intensity, however, can cause abrupt changes in population structure (Stohlgren and Rundel, 1986).

Unlike these chamise populations, non-sprouting serotinous fynbos Proteaceae have very variable population sizes from one fire to the next (Figure 4.1). Densities of mature stands (aged 15–25 years) varied over four orders of magnitude with 'typical' densities of 2000–12000 stems ranging to in excess of 80000 plants ha^{-1}. This suggests little regulation by density-dependent thinning (see also Le Maitre and Midgley, 1992). However, as we have noted above, there is evidence for density-regulated fecundity.

Bond *et al.* (1984) examined post-fire census data of serotinous Proteaceae for evidence of density-dependent population regulation. They compared post-burn and pre-burn densities in a number of subpopulations burnt in the same fire. In several cases there was a significant linear or near linear relationship between λ (measured as seedlings/parent skeletons) and pre-burn populations, especially those burnt in autumn when seedling recruitment was highest. This allows us to use our simple version of the discrete logistic equation to generate population trajectories. The result, for three populations, is shown in Figure 4.17. The predicted growth of the populations is very different. One population of *Protea repens* (from a more arid region) showed damped oscillations approaching stable equilibrial densities through successive fires. A second population from a more mesic site showed stable oscillations from low to high densities with successive fires. In the case of *Protea lorifolia*, chaotic dynamics would be predicted with a density-induced population crash within a few fire cycles. The decline to extinction is an artefact of the model and other forms of the logistic would not produce extinction (e.g. Bellows, 1981).

A possible biological explanation for oscillatory dynamics in these serotinous Proteaceae is that plants are so crowded at very high densities that they fail to reproduce – there is overcompensation (Figure 4.16). If this is the case, a decline in post-burn numbers should be observed where pre-burn densities are very high. Bond *et al.* (1984) reported one fire where seedling densities declined, for otherwise inexplicable reasons, to 12% of their pre-burn values of *c.* 80000 plants ha^{-1}. It is also possible that chaotic

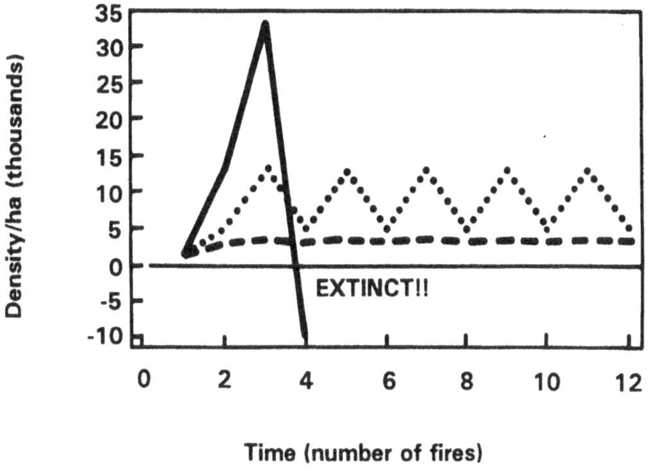

Figure 4.17 Population projections for serotinous Proteaceae based on the discrete logistic equation. Parameters for the discrete logistic equation were calculated from post-fire census data reported by Bond *et al.* (1984). Note that other growth models with density dependence would not produce extinction (e.g. Bellows, 1981). Solid line, *Protea lorifolia*; dotted line, *Protea repens* (mesic site); dashed line, *P. repens* (xeric site).

dynamics are an artefact of the model and its assumptions (e.g. low interfire mortality) and the limitations of the available data (spatial rather than temporal series). However, this example illustrates that at least a part of the variability in obligate non-sprouter shrub dynamics (Figure 4.1) might be attributable to density-dependent regulation.

Fluctuating populations are usually attributed to fluctuating environmental conditions, and especially different fire properties. What if they were caused by density-dependent regulation? How does one distinguish between the stochastic and deterministic alternatives? Density-driven oscillations are likely to be rare in most plant species because of the stabilizing effect of self-thinning and the weak relationship between seeds and established populations. Are serotinous species in nutrient-poor environments an exception?

Finally, these models suggest that local extinction may not be infrequent in non-sprouting shrubs. Field observation suggests this is true, but the causes are various (e.g. fire cycles too short or in the 'wrong' season, Chapter 5). If local extinction does occur, we will have to return to our original general demographic equation:

$N(\text{now}) = N(\text{then}) + \mathbf{B}\text{irths} - \mathbf{D}\text{eaths} + \mathbf{I}\text{mmigrants} - \mathbf{E}\text{migrants}$

and re-examine the magnitude and implications of **I** and **E** (dispersal) for these populations.

4.5.4 Grasses and self-induced population cycling

Recently, Tilman and Wedin (1991) have shown that perennial grasses may also exhibit chaotic dynamics. As discussed earlier, grass growth is often inhibited by self-shading from litter. Without fire or grazing, litter may accumulate to the extent that the population crashes. Tilman and Wedin (1991) found evidence for just such self-induced oscillations in planted monocultures of *Agrostis scabra*. Populations growing on unproductive, low-nitrogen soils maintained fairly stable above-ground biomass, those on richer soils oscillated more, and those on the richest soil showed a 6000-fold crash. These results suggest that *Agrostis* biomass underwent oscillations of increasing amplitude, leading to chaos, as productivity increased. Tilman and Wedin attributed the pattern to inhibition of growth by accumulated litter.

The mechanism of self-suppression by litter, which drives the oscillations in biomass, has been recognized in many productive grasslands. The species most likely to be affected will be those intolerant of shading (such as the 'decreasers' of Table 4.3). Species that can tolerate shading and penetrate a deep litter layer (the 'increasers' of Table 4.3) should show more stable biomass dynamics. In mixed stands, they should eventually displace species such as *Agrostis*. Both fire and grazing, by removing excess litter, would tend to reduce the likelihood of chaotic dynamics in native grasslands.

Tilman and Wedin's (1991) study offers interesting new insights into grassland dynamics. These insights would not have been gained from matrix models of tiller growth with their inbuilt assumption of unlimited growth. There is still much to learn about density-dependent biomass changes in grasslands.

4.6 EVENT-DEPENDENT DYNAMICS AND POPULATION REGULATION

We began this chapter by considering alternative explanations for population regulation. The fire-frequency hypothesis assumes that population dynamics is the outcome of relatively predictable patterns of survival, growth and reproduction periodically reset by a fire. Yet sometimes fire has catastrophic effects which are not easily explained by fire frequency. Density-dependent regulation may explain an as yet unknown proportion of these catastrophes. But fire varies in other attributes besides frequency and these may have large effects.

Bradstock and O'Connell (1988) and Bradstock (1990) have modelled the dynamics of *Banksia* using life tables of fecundity, mortality and seedbank decay rates. They used their models to predict the effects of fire frequency on population size. However, fire season and fire intensity were far more important in determining seedling numbers. Spring fires or low-intensity fires had an order of magnitude fewer seedlings than those predicted for early winter or high-intensity fires. Even in chaparral with its high seedling densities, *Ceanothus* populations have been reported to collapse after winter burns (Horton and Kraebel, 1955; Parker and Kelly, 1989). In these examples, interval-dependent effects are often of lesser magnitude than event-dependent effects in regulating plant populations.

We have already cited the catastrophic effects of intense fires for *Widdringtonia* populations, which again disrupt the predictions of 'clockwork' demographic models (Box 4.1). Further examples will be discussed in Chapter 5. The point is that each fire is a unique event and the effects of the event are less predictable than the effects of fire return interval. This can lead to very irregular population dynamics and great difficulties in prediction. Ironically, far more is known about such environmentally driven population fluctuations than those predicted by theoreticians for strong density-dependent regulation.

4.7 CONCLUSION

There is a surprising wealth of data on plant demography and fire. We know patterns of interfire survival and seed production for a number of trees and shrubs and there are even a few comprehensive studies of seedbank dynamics. Most of these studies have been directed at the management of populations by controlling fire regime. Their value in addressing conceptual issues in ecological and evolutionary studies of long-lived plants has barely been considered. Far fewer demographic studies are available from grasslands and savanna. However, these areas have a long tradition of experimental studies on the effects of fire frequency and season (Chapters 5 and 6) which could be used to test demographic models.

How successful are demographic analyses for predicting population dynamics? By far the majority of studies have been descriptive. Where predictive models have been produced, the predictions have seldom been tested against observed field studies. For example, seedbank studies of serotinous *Banksia* species predict increasing recruitment with increasing fire interval but there are no quantitative tests of the predictions. There are good reasons for the difficulty of prediction. Fire life-histories are complex. It is not enough to have detailed life-history information on interfire survival and fecundity. Seedbank dynamics must also be known and linked, somehow, to the dynamics of the plants that produce them. The implications of different combinations of $\Sigma l_x m_x$, seedbank survivorship curves and disturbance frequencies have yet to be explored. Even if interfire dynamics are well understood, the nature of the fire itself and the immediate post-fire conditions can unscramble pre-burn predictions. Finally, the difficult problem of density-dependent regulation with its apparent potential to generate chaotic dynamics awaits critical analysis.

5 Plant demography and fire II. Event-dependent effects

5.1 INTRODUCTION

In Chapter 4, we discussed the view that vegetation dynamics in fire-prone systems are essentially driven by demographic processes between fires. Fire is important merely in terminating these processes – it freezes the 'clock' at a particular 'hour' which is the start of the next fire cycle. Constant fire cycles should thus generate constant population densities and constant community composition. This is an equilibrial view of fire which assumes that all fires have equal effects. But all fires are not equal. Fire effects can vary enormously depending on the particular circumstances of a burn. These **event-dependent** effects undermine the use of demographic models (Chapter 4) for predicting fire responses. In this chapter we focus on the importance of fire events in regulating populations as opposed to the **interval-dependent** processes discussed in Chapter 4. Event-dependent effects become important when fires at the same interval produce significantly different population responses (Figure 5.1).

There are three main aspects of the fire regime which can vary independently of fire frequency. These are fire intensity (including fire type), season of occurrence and burn area (Chapter 2). All can be modified to a greater or lesser degree by human intervention. Prescribed burns are nearly always constrained by the risk of damage to people and property (Chapter 8). These can produce very different seasons and intensities from the wildfires under which natural vegetation patterns evolved. Climatic conditions both before and after a fire can also have pronounced effects on recovery and add to the stochastic nature of fire effects. The relative importance of these event-dependent factors varies among and within different vegetation types.

We first discuss the magnitude of event-dependent effects for different fire life-histories and different vegetation types (sections 5.2–5.6). We then discuss models that attempt to explain them (sections 5.7–5.9). These are quite different from models used for interval-dependent effects and often require physiological or morphological information. These explanatory models form the basis for identifying vital attributes which are helpful in

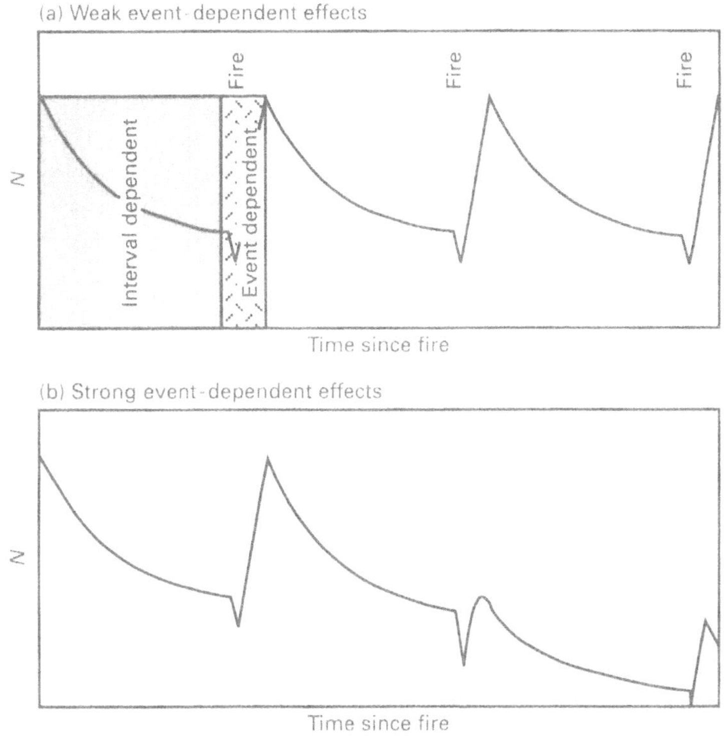

Figure 5.1 The distinction between interval-dependent and event-dependent population regulation. Interval-dependent regulation depends on processes controlling mortality and fecundity between fires. Event-dependent regulation depends on the effects of a particular fire on mortality and regeneration. The diagram illustrates a fire-recruiting sprouter life-history (Chapter 4).

predicting plant responses to different fire seasons or intensities (section 5.10). A similar approach for fire-interval effects, based on life-history attributes, was discussed in Chapter 4 (Noble and Slatyer, 1980). Finally, we note that the magnitude of fire-intensity and fire-season effects varies greatly among different biomes which, in turn, affects community stability.

5.2 EFFECTS OF FIRE INTENSITY AND SEASON ON SURVIVAL

5.2.1 Non-sprouters

The extent to which the fire event affects a population depends strongly on its fire life-history. Non-sprouters, especially trees, can survive low-intensity burns but are killed if crown fires develop. Variability in survival depends both on plant size and the range of fire intensities. In shrublands, canopies of most plants are routinely consumed in crown fires killing all non-sprouters. Where the vegetation is taller, fire response is more variable. Low-intensity fires, often favoured in prescribed burns, creep along the forest floor, consuming litter and killing low saplings but otherwise causing few canopy tree deaths. Under more extreme weather or fuel conditions, a crown fire in the same community can cause extensive mortality and initiate replacement of the stand. The dynamics of conifer forests are particularly influenced by fire intensity (Heinselman, 1981). Sprouting is rare in conifers and many species are killed by injury to the cambium or scorching of the crown (Chapter 3; Peterson and Ryan, 1986; Ryan and Reinhardt, 1988).

Both size structure and species composition of forests can be altered by different fire intensities. Late successional conifers can be eliminated from a stand by a crown fire. Because they have neither soil- nor canopy-stored seedbanks, recovery may be slow and depend on seed dispersal from patches that escaped burning. Recent studies of North American conifer forests show that different mixtures of low- and high-intensity fires produce different forest composition, depending on tree life-histories (Heinselman, 1981; Bergeron and Brisson, 1990; Bergeron, 1991; Hansen *et al.*, 1991a; Chapter 7). In these conifer forests, fire intensity can vary so much that fires are obviously far from equal in their effects. Differences in fire intensity create havoc with interval-dependent models. The state of a population before burning is a poor predictor of the state after a burn where mortality depends so much on fire intensity.

5.2.2 Sprouters

There have been many studies of sprouting in relation to fire intensity and fire season, with highly variable results. Some species are strongly affected by fire intensity, others by fire season, some by neither. In a study of vegetative recovery in British heathlands (Mallik and Gimingham, 1985), several species were subjected to heating for 2 min at 400, 600 and 800°C as a proxy measure of fire intensity (Table 5.1).

Some species were killed at high temperatures (*Juniperus communis*), others sprouted vigorously at lower temperatures but recovered poorly at higher temperatures (*Arctostaphylos uva-ursi*, *Empetrum nigrum*). One species, *Succisa pratensis*, had more regrowth at the higher temperatures, while several species, including two grasses, showed similar recovery over all the temperatures used. Clearly, the effects of fire intensity on heathland composition will be highly variable among species.

Within a species, recovery after different fire intensities usually depends on plant size. Moreno and Oechel (1991b, 1993) studied the effects of fire intensity on mortality of chamise, *Adenostoma fasciculatum*, by adding fuel in an experimental chaparral burn. This species resprouts from buds in a lignotuber at or just below the soil surface. Shrub mortality increased strongly with fire intensity especially in the smallest size classes. More than 70% of the smallest individuals were killed at high fire intensities, compared

Table 5.1 Effect of fire temperature (°C) on biomass and number of resprouting shoots in British heathland species. Biomass was measured 17 months, and number of sprouts 3 months, after burning. Values are means for five plants. Temperatures were maintained for 2 minutes using a butane gas burner to simulate buring (ex Mallik and Gimingham, 1985)

Species	*Bud position*	*Biomass (g)*			*Number of sprouts*		
		400°C	*600°C*	*800°C*	*400°C*	*600°C*	*800°C*
Juniperus communis	Above ground	3	0.8	0	0	0	0
Arctostaphylos uva-ursi	Above and below ground	30	54	26	44	48	26
Erica cinerea	At or just above ground	67	51	41	121	86	74
Erica tetralix	At or below ground	85	91	62	333	375	227
Vaccinium myrtillus	Above and below ground	85	30	3	180	51	22
Vaccinium vitis-idaea	Above and below ground	19	16	14	31	33	27
Empetrum nigrum	Mostly above ground	8	15	2	29	71	1
Genista anglica	At or just above ground	9	5	6	27	19	24
Pyrola media	Below ground and leaf bases	0.3	0.6	0.5	2	1	2
Succisa pratensis	Protected by leaf bases	49	169	202	2	2	5
Deschampsia flexuosa	Protected by leaf sheaths	113	137	102			
Festuca ovina	Protected by leaf sheaths	129	138	91			

to only 30% at control intensities. Of the largest shrubs, 20% were killed at high fire intensities but none died at lower intensities. The rate of sprouting was slower after high-intensity burns, which led to further mortality from browsing pressure when herbivores returned to the burn area. Fire season also has marked effects on chamise survival. Rundel *et al.* (1987) reported only 13% mortality after an autumn burn but 75% mortality after a spring burn, with survival highest in the largest plants. These fire-dependent survival patterns limit the value of the demographic model for this species discussed in Chapter 4 (Stohlgren and Rundel, 1986).

For species such as chamise, event-dependent effects can control population growth by selectively killing the smallest size classes. The size dependence of sprouter survival has also been a central focus of studies of savanna trees. Mature trees are often very resistant to burning but fire kills young plants, helping to maintain an open parkland savanna. For example, honey mesquite (*Prosopis glandulosa*), once a scattered tree in grassy parklands, can form dense thickets in Texas plains when fires are reduced or grazing intensified. Mesquite densities can be controlled by frequent burning (Wright *et al.*, 1976). Mortality in simulated fires of different intensity was related more to plant size than intensity, except at low fire temperatures when most plants survived (Table 5.2). Wright *et al.* (1976) could match their results to known fire intensities in different grassland fuel types and thus predict the probability of killing mesquite using fire. This probability is small for mesquite, and many other savanna trees, since seedlings rapidly acquire resistance to burning. Here an interval long enough for seedlings to grow to resistant stages appears more important than fire intensity in limiting recruitment (Chapter 4).

In savannas, changes in plant architecture may be more important than fire-induced mortality. Intense fires alter the architecture of plants, causing a reduction in foliage height. The amount of foliage killed depends both on tree height and fire intensity. A common pattern in savanna trees is shown in Figure 5.2. Light fires burn only the lower foliage and the canopy regenerates from unburnt twigs or epicormic sprouting. With increased fire intensity, epicormic sprouting may be more common. In very intense burns, the canopy is destroyed ('topkill') and trees coppice from the base of the trunk.

Table 5.2 Relationship between age, fire intensity and mortality (%) of mesquite trees (*Prosopis glandulosa*) 15 months after burning. Fires of different intensity were simulated by heating individual plants to a range of temperatures (°C) using a gas burner. Means within a row followed by the same letter are not significantly different ($P > 0.05$) (from Wright *et al.*, 1976)

Age (years)	*Temperature (°C)*				*Control*
	120	*240*	*430*	*615*	
0.5	43 b	91 a	100 a	100 a	14 c
1.5	60 b	100 a	100 a	100 a	0 c
2.5	20 c	40 a	64 a	72 a	0 d
3.5	8 a	8 a	8 a	8 a	4 a
10	0 a	0 a	4 a	8 a	0 a

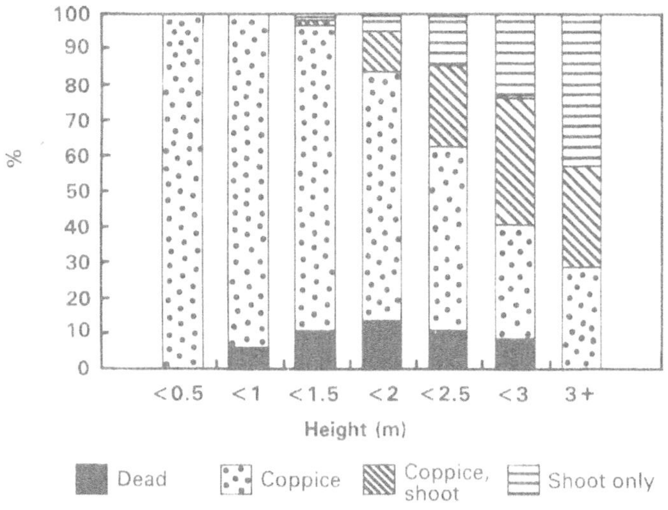

Figure 5.2 The relationship between post-burn sprouting behaviour and size in a savanna tree (*Acacia karoo*). The frequency of trees in each size class which died, coppiced from the base, coppiced from both base and canopy shoots or from canopy shoots only is shown. Sample size from smallest to largest size class was 14, 117, 128, 132, 99, 59 and 7 trees. (Data from Trollope, 1984.)

These architectural changes, induced by different fire intensities and varying with tree size, expose the sprouting shoots to increased herbivory and suppression by grasses (Trollope, 1984; Menaut *et al.*, 1990). The cost of replacing burnt foliage from stored reserves may also lead to mortality (see below). In South Africa, tree densities in rangelands have been controlled by using fire, to initiate basal coppicing, and goats, to utilize the browse that then becomes available (Trollope, 1984). These indirect effects of fire and browsing may be generally important in maintaining savannas (Chapter 7).

Fire season is also important in savannas. Many long-term burning experiments have tested the effects of both fire season and frequency on the grass/tree balance. One of the earliest experimental studies tested seasonal effects on grass/tree mixtures in Central African woodlands (Trapnell, 1959). Three treatments were compared – annual burns early in the dry season, annual burns late in the dry season and no burns. The last permitted massive recruitment of tree seedlings, initiating a successional trend from savanna to forest. After only 11 years of annual burning, the original woodland had been converted to near-grassland after the end-of-dry-season burns. In contrast, early dry-season burns led to a much denser woodland, due to extensive tree recruitment (Tables 5.3 and 5.4). These fire-season effects produced large population changes even though fire intervals were held constant. Demographic studies which ignored season effects would fail to predict these major shifts in community composition.

Table 5.3 Effects of fire season on tree mortality after 11 years of burning in Zambian woodlands. Two experimental plots (40 × 100 m) were burnt annually early (June–July) and late (October) in the dry season. A third was left unburnt. Mortality was calculated only from trees known to be present in 1933 (initial number). (Data ex Trapnell, 1959)

Treatment	*Initial number*	*Losses*	*Mortality (%)*
Late dry-season burn	161	28	17.4
Early dry-season burn	130	9	7.0
Unburnt	192	8	4.2

Table 5.4 Effects of fire season on tree densities in 40 × 100 m plots of common canopy and understorey trees in Zambian woodlands after 11 years of annual burning. 1933 values are counts of trees >15.2 cm circumference at breast height. 1944 values are counts of trees >4.6 m in height. 1944 values for saplings are counts of small trees from 1.1 to 4.6 m in height. See Table 5.3 for fire treatment. (Data ex Trapnell, 1959)

Species	*Late dry season*	*Early dry season*	*Unburnt*
Julbernardia paniculata			
1933	27	23	39
1944, trees	22	23	34
1944, saplings	2	30	112
Isoberlinia angolensis			
1933	19	17	8
1944, trees	13	16	6
1944, saplings	1	10	22
Brachystegia spiciformis			
1933	8	3	4
1944, trees	5	3	6
1944, saplings	0	18	44

Similar results have been reported from experiments in West and southern African savanna (Rose-Innes, 1972; van Rensburg, 1972; Brookman-Amissah *et al.*, 1980). However, there are exceptions. In northern Australian savannas, seasonal effects were difficult to detect after more than 10 years of annual burning (Bowman *et al.*, 1988) and the same is true of *Baikea* savanna in Africa (Geldenhuys, 1977). Both of these studies emphasized the importance of edaphic factors, rather than fire, as determinants of grass/tree mixtures. Possible reasons for these discrepancies will be discussed further below (section 5.8.4).

Fire season is particularly important for the recovery of species which spread clonally by root or stem suckering. Clonal shrubs are common in the understorey of temperate deciduous and conifer forests where they suppress recruitment of canopy trees. These species generally have buds too deeply buried to be influenced by high-intensity burns. However, their resprouting vigour can be affected by fire season. For example, the density of stems of hazel (*Corylus* spp.) increased nearly fivefold with four successive spring burns, but was reduced to half the unburnt densities after the same number of summer burns (Buckman, 1964). Harrington (1985) reported similar results for gambel oak (*Quercus gambelii*), an understorey species in Ponderosa pine forests. Cover declined with successive summer burns but increased significantly after spring and autumn burns.

The length of the fire interval also influences clonal species, partly through the thinning of stems as the population of ramets ages. The relative magnitudes of event- and interval-dependent effects on recovery have been studied experimentally in *Quercus coccifera* in French garrigue, a Mediterranean-type shrubland (Malanson and Trabaud, 1988). Here variation in fire season proved to have much greater effects on short-term oak recovery than either fire intensity or stand age (Table 5.5).

5.2.3 Grasslands

Grass recovery from burning is often strongly affected by fire season. Some of the earliest and longest-running field experiments in ecology explored the effects of fire season on the species composition of grasslands. One of the oldest was initiated in Kansas tallgrass prairies in 1926 (Towne and Owensby, 1984). Here fire season was varied but frequency held constant. Late spring burning led to a doubling of biomass in big bluestem, *Andropogon gerardii*, compared to fires burnt a few weeks earlier (Figure 5.3). The conspecific *A. scoparius* responded in the opposite manner, benefiting from earlier burns. The composition of other grass species, sedges and forbs were also significantly affected by the season of burn (Towne and Owensby, 1984).

A similar study in South African tall grassveld tested the effects of both fire season and fire frequency (Le Roux, 1989). Grass species responded strongly to variation in fire interval. One of the dominants, *Themeda triandra*, increased after short return intervals while another, *Tristachya leucothrix*, increased with longer intervals (Figure 5.4). The same two species also responded differently to fire season. *Themeda* cover declined after autumn burns at the end of the growing season relative to winter (dry

Table 5.5 Resprouting vigour in *Quercus coccifera*, as measured by mean stem density and stem height (mm) 8 weeks after burning, in relation to season, stand age and fire intensity following experimental heating with a propane torch in French garrigue (Malanson and Trabaud, 1988). Values with different letters were significantly different ($P < 0.05$). None of the possible interactions among the groups was significant

Treatment	*Density*	*Heights*
Season		
Spring	41.4 a	625.1 a
Autumn	8.3 b	37.6 b
Years since last burn		
3	23.0 c	256.7 c
9	29.1 c	409.0 d
33	22.4 c	328.3 cd
Fire intensity (minutes' duration)		
Low (1)	24.2 ef	356.0 e
Medium (2)	29.7 e	354.8 e
High (4)	20.6 f	283.3 e

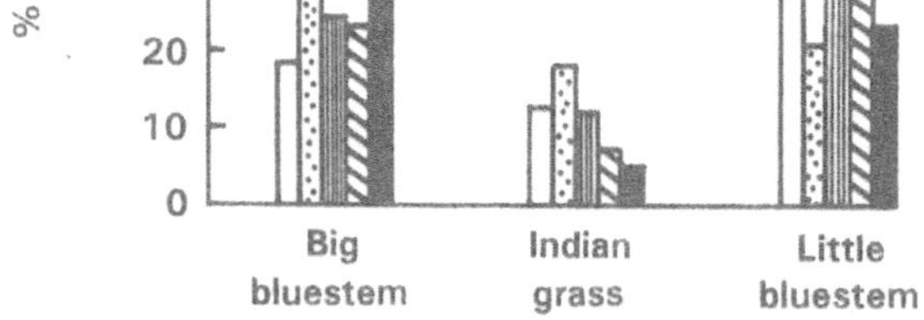

Figure 5.3 The effects of fire season on Kansas prairie composition burnt annually from 1928 to 1982. Average percentage composition is indicated for three grass species: big bluestem (*Andropogon gerardii*), little bluestem (*Schizachyrium scoparium*) and Indian grass (*Sorghastrum nutans*). (Data from Towne and Owensby, 1984.)

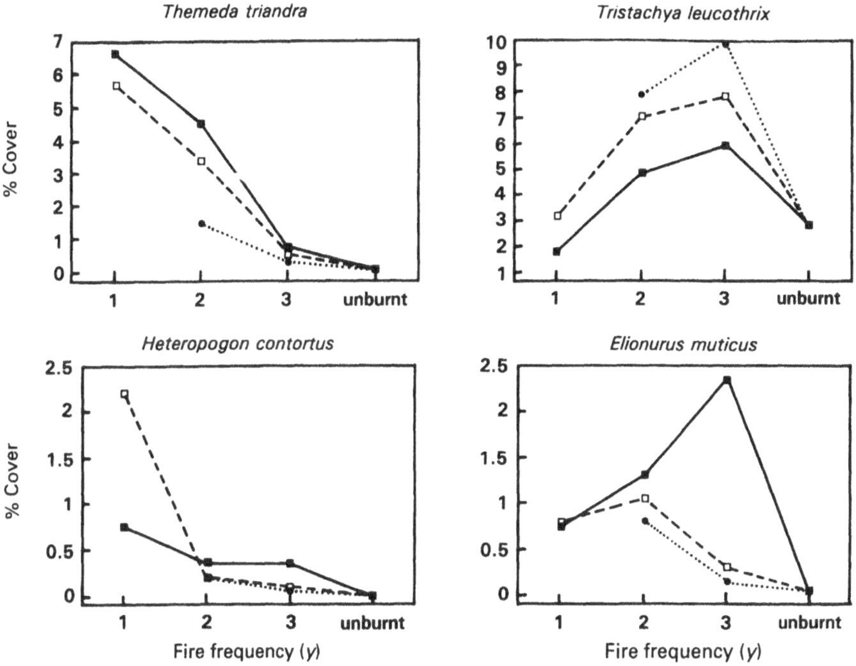

Figure 5.4 The effect of fire frequency and fire season (■, winter; □, spring; ●, autumn) on percentage basal cover of grasses in ungrazed Natal grassveld. Treatments were replicated four times in 27.4 m × 36.6 m plots and maintained for 32 years. (Data from Le Roux, 1989, Table 2.)

season) or spring (early growing season) burns. *Tristachya*, in contrast, increased after autumn burns and decreased with winter burns (Figure 5.4). In this experiment, populations responded as strongly to fire season as to fire frequency. Note that the matrix model of grass response to fire discussed in Chapter 4 had no provision for fire season and implicitly assumed that fire season was held constant (Silva *et al.*, 1991).

Grassland fire intensities are less variable than those of woody communities (Chapter 2) and there have been few studies of intensity effects. Among the most extreme differences in fire intensity are those caused by the type of burn. Backfires (against the wind) have slow rates of spread and release heat energy close to the ground, whereas headfires (burning with the wind) are faster and hotter higher in the vegetation (Trollope, 1978). Bidwell *et al.* (1990) studied the effects of fire type in tallgrass prairies using replicated fires in the same grass community. The headfires spread more than 10 times faster than backfires (7.5 v. 0.6 m min^{-1}). Headfires produced 21% more tallgrass standing crop than backfires over a growing season, but 26% less forb phytomass. Some grasses with a different growth form, such as little bluestem (*Shizachyrium scoparium*), were unaffected by the fire treatments. Similar results were reported by Trollope (1978, 1982) for African grassland and savanna, with headfires significantly favouring tallgrasses and causing extensive topkill of *Acacia karoo* trees, whereas the trees were unaffected by backfires. Trollope attributed the different responses to the distribution of heat energy in relation to location of buds. Both studies suggest that fire intensity can have significant effects on grassland composition unrelated to fire frequency. However, variation in fire intensity in grasslands is probably too small to be a significant factor at a landscape scale.

5.3 EFFECTS OF FIRE INTENSITY AND SEASON ON RECRUITMENT

The fire event influences not only survival but also recruitment, especially in fire-recruiting species. The most important effects are among non-sprouting species. These life histories are most prominent in fire-prone shrublands where the effects of fire intensity and season have been most studied. Here too, event-dependent effects have proved as, or more, important than interval-dependent phenomena in driving population changes.

5.3.1 Fire intensity

Our first example is an experimental study of the effects of fire intensity on chaparral regeneration (Moreno and Oechel, 1991a). Fuel was added as cut brush before a chaparral burn to simulate the effects of different fire intensities. The two dominant shrubs, *Adenostoma fasciculatum* (chamise) and *Ceanothus greggii*, and the post-fire herb flora responded in diverse ways to fire intensity (Figure 5.5). Chamise recruitment was reduced at high fire intensity whereas *Ceanothus* recruitment increased or remained constant. Populations of annuals all declined with increasing intensity. However, the fire-recruiting annual, *Phacelia brachyloba*, was less affected by high intensities than the opportunistic annual, *Camissonia intermedia*. Another fire-recruiting species, *Lotus strigosus*, a short-lived shrub, showed highest seedling emergence at the highest fire intensities.

Hard-seeded legumes such as *Lotus* seem to require high fire temperatures for germination. In Western Australian jarrah (*Eucalyptus marginata*) forests, seeds of understorey *Acacia* spp. failed to germinate in low-intensity management burns. The resulting low density of these legumes not only reduced nitrogen input into the system but also led to the spread of the pathogenic fungus, *Phytophthora cinnamomi*, which causes dieback of the valuable jarrah trees (Shea *et al.*, 1979).

In conifer forests, variation in fire intensity can alter recruitment conditions for canopy trees by altering soil surface conditions. Intense fires remove the duff layer of fallen leaves so favouring recruitment of species which require a mineral seedbed for germination. Low-intensity fires, in contrast, promote recruitment of larger-seeded species which can tolerate the duff layer (Heinselman, 1981).

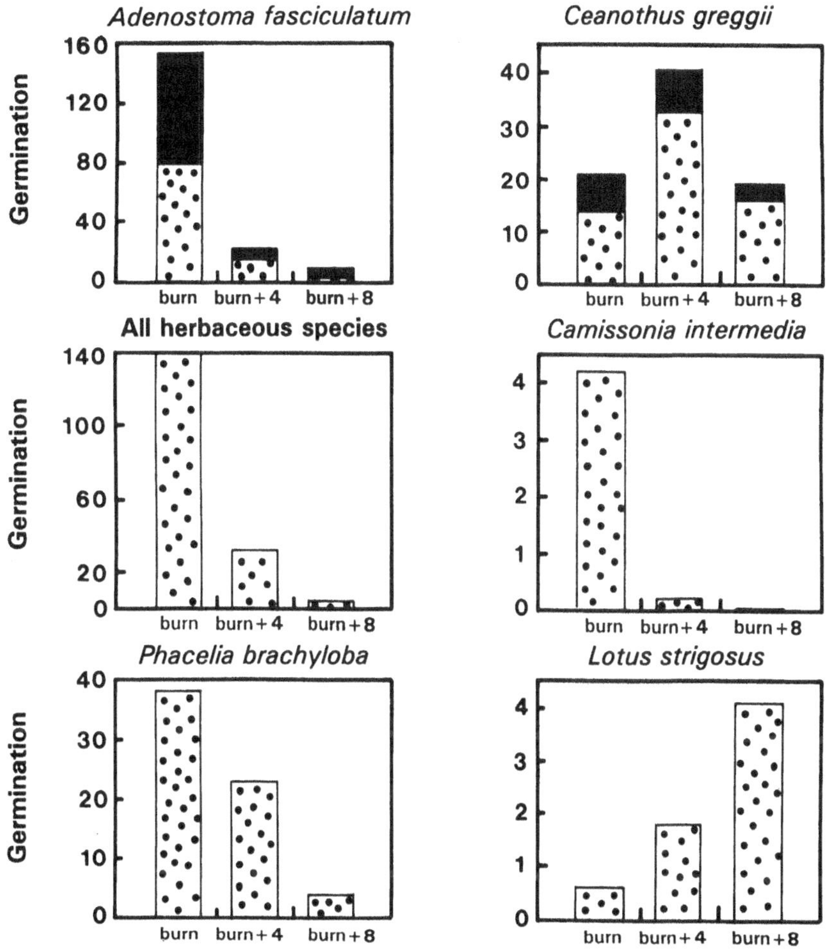

Figure 5.5 Fire intensity and germination responses (number of seedlings m^{-2}) in Californian chaparral. Three levels of fire intensity were simulated by adding fuel: burn, no fuel additions; burn + 4, 4 kg m^{-2} cut brush added; burn + 8, 8 kg m^{-2} added. *Adenostoma fasciculatum* is a sprouter and *Ceanothus greggii* a non-sprouting shrub. *Lotus strigosus* is a short-lived shrub. Both *Phacelia brachyloba* and *Camissonia intermedia* are annuals, the former with fire-stimulated germination and the latter without. Seedlings emerged in the first (dotted) and second (black) year after the burn. (From Moreno and Oechel, 1991a.)

These and similar studies (see below) show that fire intensity can directly influence community composition by selectively influencing recruitment. Thus in chaparral, low-intensity fires would tend to favour chamise and opportunistic annuals, whereas high-intensity fires would favour non-sprouting *Ceanothus* and the fire-recruiting annuals and pauciennials (Moreno and Oechel, 1991a).

5.3.2 Fire season

Fire season can also have pronounced effects on recruitment. In South African fynbos, winter (wet-season) and spring (early dry-season) burns caused population crashes of serotinous Proteaceae. Densities were reduced tenfold after some spring burns. In contrast, autumn (late dry-season) burns led to heavy seedling recruitment (Bond *et al.*, 1984; Figure 5.6).

Similar seasonal effects have been reported in Australian *Banksia* species, which also decreased after spring burns (Cowling and Lamont, 1987; Bradstock and O'Connell, 1988; Bradstock and Bedward, 1992).

In chaparral, chamise (*Adenostoma fasciculatum*) recruitment can be strongly affected by fire season. Rundel *et al.* (1987) reported low seedling densities after an autumn burn ($17\,m^{-2}$ with only $0.5\,m^{-2}$ surviving after 2 years) compared with a spring burn ($67\,m^{-2}$ with more than $4\,m^{-2}$ surviving after 2 years). In contrast, Horton and Kraebel (1954) and Parker and Kelly (1989) reported marked reductions in non-sprouting *Ceanothus* populations after spring burns.

In some herbaceous plants, fire season affects not only seed biology but also fire-stimulated flowering. For example, fire season strongly affected flowering in the fynbos geophyte, *Watsonia borbonica* (Le Maitre and Brown, 1992). Flowering was stimulated by autumn burns at the end of the summer drought, when the species produces masses of pink flowers, but was strongly inhibited by spring burns (Table 5.6). Fruit set was comparable in both fire seasons (45.7% autumn, 36.4% spring) but seed release was much greater in the autumn-burnt populations, probably due to predator satiation in the mass flowering populations (9.8% seed predation after autumn burns v. 56.6% after spring burns). The net result of one autumn burn was

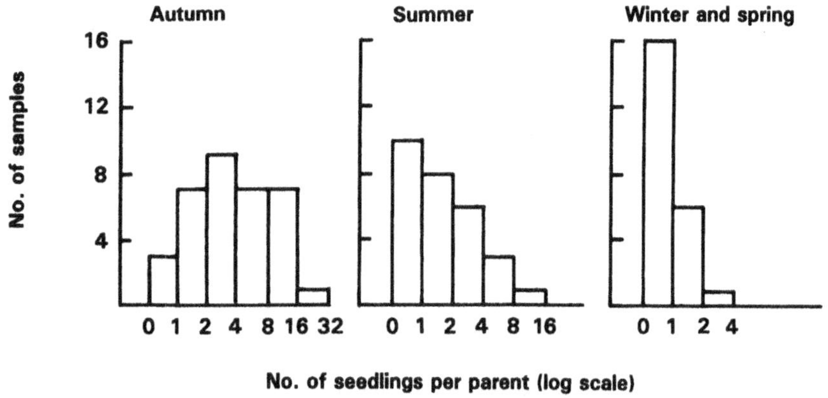

Figure 5.6 Variation in recruitment associated with fire season in serotinous fynbos Proteaceae (Bond *et al.*, 1984). All the species were non-sprouters. Seedlings were censused after burns, along with burnt skeletons of the pre-burn population.

Table 5.6 Flowering response of *Watsonia borbonica* after an autumn and a spring fire in Cape fynbos (Le Maitre and Brown, 1992). Successive censuses showed that flowering was fire-stimulated in these populations

Year of assessment	*% in flower (sample size in parentheses)*	
	Autumn 1975 fire	*Spring 1975 fire*
1975	49.9 (485)	8.9 (248)
1976	0 (615)	<0.1 (248)
1977	<0.1 (609)	<0.1 (255)

a 7500% increase in seedling recruitment but there was a 0% increase in a nearby spring burn (Kruger, 1983).

5.4 EFFECTS OF CLIMATE VARIATION ON RECRUITMENT

Where fires initiate a wave of seedling recruitment, post-burn weather conditions can influence recruitment success. Drought has been particularly emphasized in Australian studies. Lamont *et al.* (1991b) found that population fluctuations of the rare *Banksia cuneata* were driven not by the size and dynamics of the canopy seedbank but by drought-induced mortality of seedlings immediately after a fire. Also in Australia, Hodgkinson (1991) showed large differences in understorey shrub recruitment in Australian semi-arid woodlands correlated with rainfall after the burn. Bradstock and Bedward (1992) simulated shrub recruitment from climatic data in eastern Australia to compare the relative importance of fire season and rainfall variation in driving population oscillations in serotinous shrubs. Over their 50 year simulation, the timing of fire relative to sequences of wet and dry years was as important as fire season in regulating *Banksia* populations.

Burgman and Lamont (1992) simulated extinction risks in *Banksia cuneata* to explore the viability of populations under different managemnt options. Their simulation showed that the greatest risk of extinction was due to recruitment failure in drought years. Short fire intervals increased the risk of extinction because populations would experience more recruitment events, some of which might be drought years. Burgman and Lamont's model is an elegant example of a new generation of demographic models. It combines both deterministic fire-interval and stochastic event-dependent processes for predicting the probability of persistence, and not only density, for a fire-prone population (Box 5.1).

Box 5.1 Modelling population viability for *Banksia cuneata*

Burgman and Lamont (1992) developed a population model for *Banksia cuneata*, a rare, non-sprouting serotinous shrub. The model combined deterministic effects of fire interval on survival and reproduction and stochastic effects of drought on seedling recruitment. Survival was modelled as age-dependent attrition of a cohort of seedlings establishing after fire until death at 45 years. Interval-dependent recruitment was modelled by incorporating a juvenile period lasting 4 years, followed by constant low seed production until 12 years and then constant high seed production until death. Recruitment was modelled as a linear function of rainfall in the first year after fire, adjusted for plant age when the population burned. A stochastic element was introduced by generating recruitment from rainfall simulated to mimic natural rainfall patterns in the area. The model explored the effects of different fire intervals on population growth and extinction risk over a 50 year period. An additional stochastic element was included by simulating uncontrolled wildfires as a function of time since the last burn.

A sensitivity analysis of the deterministic version of the model indicated that rainfall was by far the most important factor determining growth of the population. The most effective management plan would be to water the seedlings in the summer drought following fire. Manipulating fire interval is a more practical option but the model indicated conflicting options between maximizing population size or minimizing extinction risk (see figure below). Long fire intervals led to

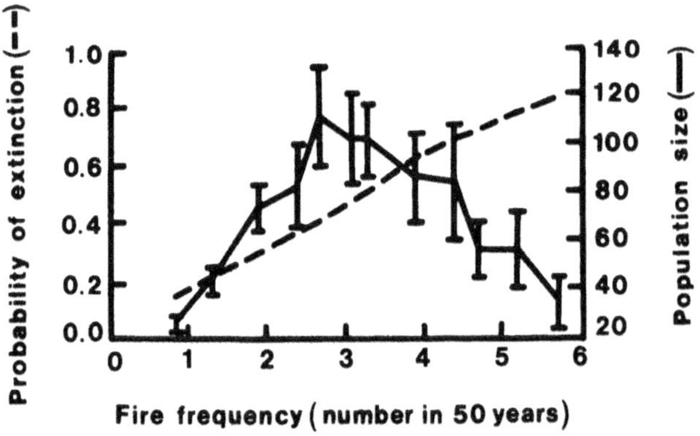

Probability of extinction (dashed line) and mean and 95% confidence limits for population size (solid line) at the end of 50 years, both as a function of mean fire frequency (number of fires per 50 years). (From Burgman and Lamont, 1992.)

(continued)

the smallest risk of extinction over a 50 year simulation because the seedlings would be exposed to fewer post-burn droughts. However, a policy of fire suppression would not work in the longer term because plants do not live longer than 45 years. Populations were more likely to grow at fire intervals of 15–25 years, but at a greater risk of extinction.

Burgman and Lamont's model illustrates nicely the value of combining interval- and event-dependent factors in models of fire-prone populations. It also shows that persistence, rather than density, may be the more useful consideration where event-dependent effects are important.

5.5 EFFECTS OF GROUND FIRES

Some of the most extreme departures from simple interval-dependent vegetation cycles are caused by 'ground fires' – fires burning in the organic layers of the soil (Chapter 2). These burn deep into peaty soils and exhibit glowing rather than flaming combustion (Radley, 1965; Wein, 1983). Since organic soils occur in wet ecosystems, ground fires are rare and restricted to extreme drought years. Typical examples of destructive ground fires occurred in the drought year of 1976 in British and French heathlands and moorlands (Clement and Touffet, 1981, 1990; Maltby *et al.*, 1990) and Canadian bogs (Wein, 1983). At Rosedale Moor, in Britain, over 600 ha of *Calluna* heathland was destroyed when fires burnt deeply into blanket peats reducing them, in extreme cases, to thick layers of ash. Both the seedbank and underground vegetative organs in the peat soils were destroyed. In a census made a decade after the fire, the original *Calluna* heaths had been transformed. Less than 6% of the area supported vascular plants, 65% was converted to bryophytes and 30% was still barren. The devastating effects of this fire were quite different from the rapid recovery after 'typical' heath burns (Hobbs *et al.*, 1984b).

Ground fires, which are extremely difficult to control, may become much more common in organic terrain in the future if global warming continues. For example, Landhausser and Wein (1993) showed large vegetation changes across a severely burnt forest–tundra ecotone in Canada. *Betula papyrifera* and *Populus balsamifera*, both deciduous trees, invaded burned tundra, replacing the previously treeless cover dominated by low shrubs. Both species are capable of long-range seed dispersal in contrast to *Picea* species, and are more persistent because of their capacity to sprout after fire. Landhausser and Wein predicted that the treeline in the western Canadian Arctic would move northwards with each successive ground fire, exposing mineral soil and providing seedbeds conducive to deciduous tree establishment.

5.6 EFFECTS OF FIRE AREA

The least-studied effect of a fire event is the size of the burn area. Burn area is important for at least two reasons. Firstly, many experimental burns are very small compared to natural or prescribed fires. How representative are these tiny burnt plots of the situations they are supposed to simulate? How do edge effects modify the results? Secondly, many natural areas have become fragmented so that the scale of landscape fires has become greatly diminished. What effect will smaller burn areas have on vegetation recovery?

There seem to have been few explicit studies of the effects of fire area on post-burn recovery. The most general effect is probably on levels of herbivory. Burnt areas attract large mammals, birds and insects to the succulent regrowth – a principle well known to stock farmers and hunters (Wright and Bailey, 1982; Bigalke and Willan, 1984; Main, 1986). Concentrations of large grazing mammals are often common in burnt grasslands or savanna. However, burning also removes cover and discourages small mammals dependent on cover to escape predators (Wright and Bailey, 1982; Bigalke and Willan, 1984). Herbivory levels are likely to be much higher in small burns either because animals congregate on these small, palatable islands or because food sources in the burnt area are close to unburnt cover.

High levels of small mammal herbivory have sometimes been reported in burnt shrublands. In chaparral, Mills (1983, 1986) found differential predation of *Ceanothus greggii* and *Adenostoma fasciculatum* seedlings by small mammals, especially brush rabbits (Table 5.7). He argued that the competitive advantage of *Ceanothus* seedlings was reversed by selective

Table 5.7 Mortality (%) due to herbivory or 'drought' for tagged seedlings of chamise (*Adenostoma fasciculatum*) and ceanothus (*Ceanothus greggii*). Separate cohorts were marked each year for the first 3 years after a small (2.5 ha) chaparral burn (ex Mills, 1986)

Years post-fire	*Source of mortality*	*Chamise*	*Ceanothus*
1	Mammal browse	31	52
	Drought	38	25
2	Mammal browse	1	11
	Drought	6	3
3	Mammal browse	0	20
	Psyllids	0	4
	Drought	11	0
Cumulative	Herbivory	31.3	59.1
	Drought	43.0	25.6
Total mortality		74.3	84.8

herbivory on this species. However Mills' study was in a burnt area of 2.5 ha in a sea of chaparral unburnt for several decades. This is at least 100 times smaller than typical burns in this area. As noted by Mills (1986), seedlings were all within 30 m of the perimeter of the burn and therefore accessible to the 'shy' brush rabbits which do not stray far from unburnt cover. One would expect herbivory to tail off rapidly in larger burnt areas with increasing distances of seedlings from the fire perimeter. There are several similar examples of high levels of shrub seedling predation by small mammals, usually from small experimental plots (Le Maitre, 1988) or small burns (Cowling and Lamont, 1987), sometimes contrasting sharply with much lower levels recorded in larger burnt areas (Midgley, 1988).

Calluna heathlands in Scotland provide an interesting example of the importance of burn area (Hobbs and Gimingham, 1987). Red grouse (*Lagopus lagopus scoticus*) feed on young *Calluna* stands but return to older stands for nesting and for protection from predators. These conflicting requirements are met by burning narrow strips of heathlands to promote grouse populations (Chapter 8). Heather burns to promote regrowth for feeding sheep are much larger (10 ha or more).

It seems likely that area effects will be important in determining levels of seedling herbivory in woody vegetation. However, with the exception of the grouse example, we have very little idea of the shape of the curve relating burn area to herbivory or the possible existence of thresholds above which herbivory is spread too thinly to be significant.

Burn area also influences the chances of seed dispersal from adjacent unburnt populations. This may be a particularly significant factor in recovery of conifer forests after fire, especially where the trees are not serotinous.

5.7 THE CAUSES OF EVENT-DEPENDENT EFFECTS

As outlined above, many studies have demonstrated strong event-dependent effects of fire. Three important points emerge:

- variation in fire intensity and fire season are important, and often independent, determinants of population change;
- both survival and recruitment responses vary between species burnt in the same fire;
- survival responses within species change with increasing plant size.

The evidence for event-dependent effects is based on extensive, often long-term, field experiments. In grasslands and savannas, some experiments have been continued for decades and are among the longest-running field experiments in ecology (Rothamsted excepted!). Unfortunately, they were seldom designed to explore the processes generating the observed vegetation changes. As a result it is difficult to judge how widely the results apply, and for which species or functional types.

It would be very useful to develop a set of rules from which one could make intelligent guesses of likely fire responses in different species. There are few precedents in the general ecological literature for a predictive framework for event-dependent responses. Herbivory, like fire, results in defoliation at different seasons and intensities (Crawley, 1983). However, rules for plant responses are also rather poorly developed (Crawley, 1983; Huntly, 1991). Interval-dependent effects of fire fit much more easily into the general framework of ecological models and theory (Chapter 4). Qualitative rules for fire response have been devised based on vital attributes, such as age to maturity and life span (Noble and Slatyer, 1980; Moore and Noble, 1990). What are the equivalent attributes for predicting event-dependent responses?

In the following section we review briefly the suggested mechanisms for fire season and intensity effects in populations. These form the basis for identifying a set of vital attributes for understanding event-dependent effects (section 5.10).

5.8 EXPLAINING SURVIVAL PATTERNS

5.8.1 Bud dynamics and fire intensity

Two alternative possibilities lie at the heart of almost all attempts to explain the effects of fire season or intensity on survival:

- Survival depends on the fate of regenerative buds.
- Survival depends on the resources available for recovery.

The fate of buds depends, in part, on the degree of exposure to fire heat. Buds can escape being scorched, consumed or killed by occurring above the heating zone (survival depends on height), below the heating zone (survival depends on depth below soil surface) or protected by insulating layers (survival depends on plant structures such as bark). The key attribute for predicting the fate of plants at different fire intensities would therefore be the location of bud tissues.

The resource argument emphasizes the importance of resources for replacing damaged tissue. Recovery is best when fires occur when stored reserves are at a maximum, such as during dormant growth periods, or when the growing season after a fire is long enough to re-build reserves. These alternative explanations have seldom been explicitly tested and interactions between them have been poorly explored.

Crown sprouting

For plants that recover from above-ground buds, such as many conifers, survival depends on the degree of canopy damage. Foresters have used logistic regression and discriminant analyses to predict conifer mortality from tree and fire properties. Canopy scorch depends on fire intensity, tree height and the height of the base of the crown (Figure 5.2; Chapter 3).

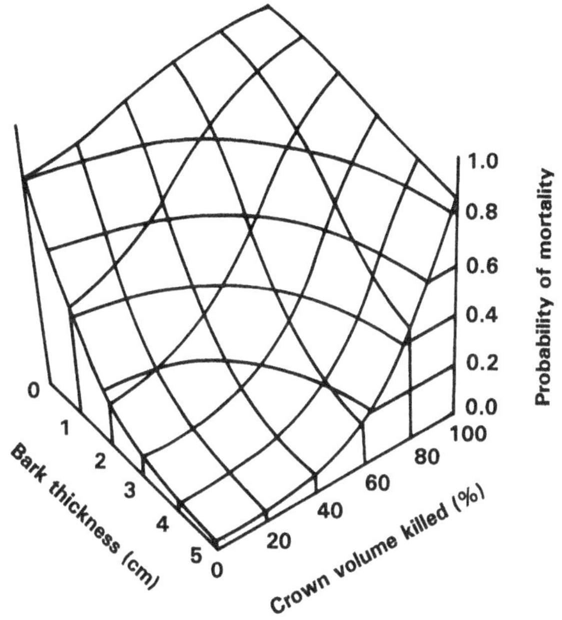

Figure 5.7 Probability of mortality of seven conifer species as a function of percentage crown volume killed by fire and bark thickness. The response surface shows that mortality is high in trees with very thin bark, regardless of crown injury. At high levels of crown kill, mortality is high even for thick-barked trees. The model was derived from logistic regression analysis applied to over 2000 trees from 43 prescribed fires in western North America. (From Ryan and Reinhardt, 1988.)

Whether the canopy recovers, depends on the amount of crown scorched, suggesting that survival is in part a resource problem, but also on the degree to which buds are shielded from heat. Among conifers, species with large buds and twigs tend to be more resistant to fire damage (Ryan and Reinhardt, 1988). Bark thickness, as discussed in Chapter 3, has been found to augment crown scorch as a predictor of conifer mortality (Figure 5.7; Peterson and Ryan, 1986; Ryan and Reinhardt, 1988) and differs between trees of different species and sizes.

Protection of the cambium by an adequate bark layer is a prerequisite for canopy survival. Damage to bark by animals is often overlooked as a factor affecting fire-induced mortality. Yeaton (1988) has shown that savanna trees damaged by porcupines were more likely to be killed by fire. Porcupines preferred the bark of some species over others. Yeaton argued that the indirect effects of porcupines on fire mortality was important in determining tree species composition.

Basal sprouting

In the first instance, fire survival from basal sprouts seems to depend on the fate of the bud bank. The state of carbon reserves is only significant for subsequent vigour of regrowth. Their relative importance has been explored experimentally by comparing the effects of cutting and burning shoots. Mortality should be equivalent in the two treatments if stored reserves were most important for recovery. Many studies have found significantly higher mortality after burning, showing that a defoliation treatment is not analogous to burning (Kayll and Gimingham, 1965; Noble, 1984; Auld, 1990; Lloret and Lopez-Soria, 1993). These results suggest that bud mortality largely accounts for the variability in recovery after different fire intensities.

A second prediction from the reserve hypothesis is that intraspecific variation in survival should depend on the size of storage organs. Most studies have found that small plants are more easily killed by cutting than larger plants. However, many more small plants are killed by fire, suggesting that survival of the bud bank is more important. Poorer survival of small plants may be due to poorer bud insulation (thinner bark, shallower roots) or fewer buds. Zammit (1988) studied the latter possibility. By repeated removal of shoots, he showed that the bud population in *Banksia oblongifolia* was proportional to plant size, but that small plants had proportionally more buds than larger plants. The importance of bud insulation for size-specific survival does not seem to have been studied (but for a first attempt see Auld, 1990).

Interspecific variation in response to fire intensity may also depend on the location of subterranean buds. For example, clonal species with deeply buried stems seem very resistant to variation in fire intensity. There has been very little study of the biology of bud tissues in fire-prone species, even in the extreme case of sprouters versus non-sprouters (Berg and Plumb, 1972). Flinn *et al.* (1993) have described a method for locating regenerative tissue that may be of general utility in interspecific comparisons. They clipped shoots of five Texan shrubs at various positions above and below the ground and observed the sprouting response. Each species showed a distinct regenerative hierarchy, with sprouting always occurring from buds closest to the cut even where the capacity for sprouting existed more distally. Three species sprouted from both stem and roots. Two failed to sprout from roots though both had lignotubers. The results suggest that the latter species would be more easily killed by intense fires. This simple experimental protocol should be very useful in ranking species for tolerance of fire intensity. It has yet to be validated in burning trials.

5.8.2 Phenology and fire season

Why should fire season influence post-burn recovery? The answer seems to lie in seasonal variation in the carbon economy. Carbon reserves vary both seasonally and from year to year (Jones and Laude, 1960; Laude *et al.*, 1961). If survival depends on adequate reserves, then the phenology of growth should indicate recovery potential (Jones and Laude, 1960). A

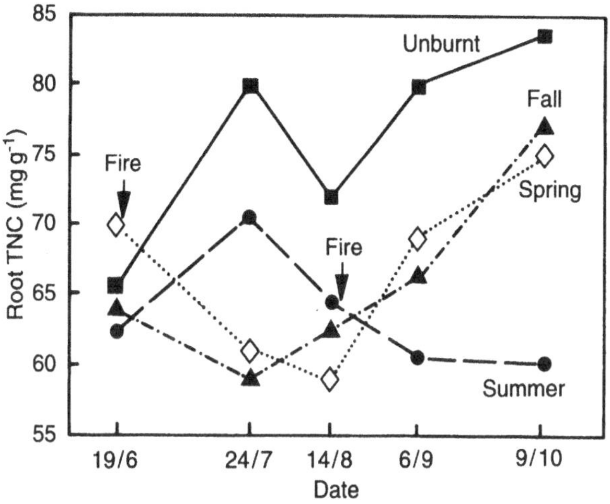

Figure 5.8 The relationship between fire season and root TNC (total non-structural carbohydrates) in gambel oak, *Quercus gambelii*. The autumn (fall) unit was burned in the previous October. Note that, following initial depletion, autumn- and spring-burn TNC reserves had recovered by winter, in contrast to the summer burn. Repeated summer burns caused a reduction in oak density, canopy cover and frequency. (From Harrington, 1989.)

number of studies have tested this hypothesis by measuring TNC (total non-structural carbohydrate) levels in roots and correlating these with the sprouting response after fire (Jones and Laude, 1960; Hough, 1968; Harrington, 1989; Figure 5.8). Rundel *et al.* (1987) interpreted the chamise responses to fire season (section 5.2.2) as being a consequence of depletion of stored carbohydrates. Roots switch from carbon source to sink in late spring, at the end of the rainy season, after initiating new shoot growth (Jones and Laude, 1960; Laude *et al.*, 1961). If a fire occurs when roots are acting as a carbon sink, then sprouting should be poor.

The correlation between recovery and stored carbon observed in these studies is not necessarily causal. The carbon reserve argument has been criticized in the grazing literature; first, because reserves have been measured by concentration and not pool size; and, secondly, because photosynthesis can rapidly replace grazed tissue (Caldwell, 1986; Richards, 1986). The importance of carbon reserves versus carbon uptake has not been tested for post-burn recovery. Fires that occur at the beginning or end of the photosynthetic season should have less effect on recovery than fires in the middle of the growing season. The latter would suffer doubly through depletion of stored reserves and a truncated growing season. Gambel oak, (*Quercus gambelii*) is a winter-deciduous species whose recovery was severely affected by summer burns, in the middle of the growing season, but not by spring or autumn burns at the beginning and end of the season (Harrington, 1989). Malanson and Trabaud (1988) attributed poor regeneration of *Quercus coccifera* after an autumn burn to the short growing season curtailed by early frosts. Unlike fire-intensity effects, seasonal effects involving depletion of reserves would only take effect over repeated fires in the 'wrong' season.

Plants that depend heavily on stored carbon to initiate new growth, such as geophytes, should be particularly severely affected by growing-season burns. In the fynbos geophyte, *Watsonia borbonica*, defoliation was less damaging when plants were dormant than when they were actively growing, presumably because stored reserves were greater in the dormant phase (Le Maitre and Brown, 1992). *Watsonia*, and presumably other fire-stimulated geophytes, may completely fail to flower if burnt in an inappropriate season, potentially delaying recruitment for decades.

5.8.3 Event-dependent effects in grasses

We know rather little about what causes vegetation changes in grassland-burning experiments. For example, it is not known whether seedling recruitment or tiller dynamics drive vegetation changes observed in long-term burning experiments (Glenn-Lewin *et al.*, 1990). Perennial grasses generally survive fire because the range of fire intensities is small and meristems are well protected by leaf bases (however, there are some reports of bud deaths due to burning in species of the genus *Stipa*; Wright, 1971; Busso *et al.*, 1993). Tainton (1981), working on South African grasslands, has suggested that the marked effects of fire season seen in many grassland-burning experiments may be caused by the degree to which growing tips are protected from heat injury. During the annual cycle of tiller initiation, growth and senescence, bunch grasses are most susceptible to defoliation after new shoots have emerged from the protective leaf bases and before they have replaced the energy and nutrient stores used for shoot elongation (Booysen *et al.*, 1963; Tainton, 1981). According to this tiller phenology hypothesis, grasses with underground stems should be less affected by burning than those that tiller from above ground. Plants whose stems remain compact, with apices protected in leaf sheaths near the ground, should be less sensitive to defoliation than those with stem apices exposed above the ground (Tainton, 1981).

Tainton (1981) has also interpreted flowering responses in terms of tiller phenology. Grass plants fail to flower and set seed if they are burnt while the culms are elongating. Grass species vary widely in seasonal patterns of development of tillers, with C_3 and C_4 grasses growing in cool and warm seasons, respectively, but with much variation within each photosynthetic pathway too. Species differ greatly in the timing and rate of flowering-stem elongation. Most species extend the flowering culm very rapidly, but some elongate slowly and will therefore be vulnerable for longer periods (Figure 5.9). There is considerable variation in tiller phenology among species, but also within species and across habitats, suggesting complex responses to fire season (Glenn-Lewin *et al.*, 1990).

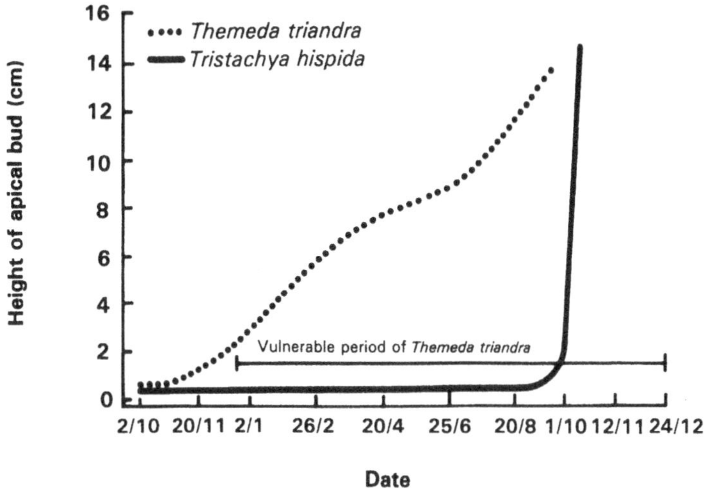

Figure 5.9 Seasonal variation in the height of the apical buds of two grasses, *Themeda triandra* and *Tristachya hispida* from Natal. The curves represent the progress of the apical buds from the time of initiation of the vegetative tillers in one season to their flowering in the next. *Themeda triandra* has buds elevated in a vulnerable position for defoliation, whether by fire or herbivory, for a much longer period than *Tristachya hispida*. (From Tainton, 1981, p. 222, Figure 8.4.)

Tainton's phenological hypothesis seems satisfyingly simple and quite general. Svejcar (1990), too, has suggested that the phenological stage of key species, rather than date, should be used to guide the use of prescribed fire in North American prairies as a tool for manipulating grass composition. However, there have been few explicit tests of the effects of defoliation at different times on flowering or tiller dynamics of co-occurring species with contrasting phenologies. Studies in American tallgrass prairies are beginning to provide the necessary detail on flowering and tillering response to defoliation at different seasons, but have yet to be linked to explicit phenological data.

5.8.4 Grass/tree interactions in savanna

How fire affects grass/tree ratios in savannas depends partly on the availability of fuel. Grass productivity varies with rainfall, soil type, grazing pressure and tree density. In high rainfall areas, such as Trapnell's (1959) Central African woodlands, grass productivity greatly exceeds losses to herbivores. Large fuel loads every dry season allow frequent intense burns with the potential to transform rapidly savanna structure. Where grass fuel loads are smaller (e.g. more arid, poorer soils, higher grazing pressure), fire intensity may be too low, in most years, to burn tree canopies or kill seedlings.

There seem to have been few attempts at explaining the results of savanna-burning experiments in terms of tree and grass biology. Changes in tree densities burnt at different seasons may depend on the phenology of carbon reserves. Deciduous savanna trees often leaf out before the rains begin. Fires at this time would presumably be particularly damaging since carbon reserves would be depleted from the production of new leaves. Grasses, on the other hand, are highly sensitive to defoliation in the growing season and can be weakened if burnt during the rainy season (Trapnell, 1959; Rose-Innes, 1972). Fire season and intensity may jointly affect grass/tree ratios. High-intensity fires, which burn more of the canopy, would result in a greater drain on resources to replace burnt foliage. Could low levels of canopy damage as a result of low grass fuel loads explain the lack of seasonal effects in Australian (Bowman *et al.*, 1988) and Caprivi (Geldenhuys, 1977) savanna experiments? Records of whether flame heights were sufficient to scorch the canopy in these burning experiments would be useful in resolving this problem.

5.9 EXPLAINING RECRUITMENT PATTERNS

Two general classes of explanation have been offered for why seedling recruitment varies after different fires:

- the effects of heating on seed germination and survival;
- the direct and indirect effects of reproductive phenology on seedbank size.

Both are analogous to explanations offered for survival of vegetative tissue. Both the number of seeds and the number of buds surviving fires depend on the heat released during a burn. Alternatively, both seed supply and the resource state of sprouting plants vary with season and pre-burn climate conditions. The alternatives have not always been effectively separated in studies of fire-stimulated recruitment.

5.9.1 Effects of fire intensity on recruitment

There is considerable variation in intra- and interspecific seedling emergence after fires of different intensities. Recent studies have gone a long way in explaining the causes of this variation in soil-stored seeds. During a fire, there is a steep gradient in temperature from the soil surface to the insulated layers below. The fate of a seed depends on its location along the temperature gradient. Seeds too near the surface will be killed. Seeds too far from the surface may germinate but be buried too deeply for seedlings to emerge. Heat-stimulated seeds can only germinate where temperatures exceed the minimum requirement but are below the lethal limits. Auld (1986, 1987) has made a very comprehensive analysis of the problem in a study of recruitment in *Acacia suaveolens*, an Australian non-sprouting shrub. This species has relatively large, myrmecochorous (ant-dispersed) seeds. Maximum seedling emergence depth was 10 cm. Seeds dispersed below this (by ants) may germinate but fail to reach the soil surface. *Acacia suaveolens* is hard-seeded and seeds required heating to between 60°C and 80°C to break dormancy (Figure 5.10).

Temperatures above 100°C were lethal, if of sufficient duration, and above 150°C even for very short duration. Auld matched soil temperature profiles recorded in simulated low- and moderate-intensity burns to temperature thresholds for survival and germination in order to predict germination percentages at each soil depth (Figure 5.11). The effects of fire intensity on recruitment could be predicted from this information coupled with data on the distribution of seeds in the soil. Results suggested that emergence following the low-intensity burn would be only one-third that of the moderate-intensity burn. Poor recruitment at low fire intensities was due to low heat penetration into the soil so that seeds failed to reach threshold temperatures for germination (Figure 5.11).

Auld and O'Connell (1991) have extended this approach to a large number of shrubby Fabaceae from eastern Australia. In general, seedling emergence would be greater in high-intensity burns because the hotter fires stimulate seed germination from a greater range of soil depths. Several field studies in Western Australia (Shea *et al.*, 1979), Californian chaparral (Moreno and Oechel, 1991a) and Cape fynbos (Bond *et al.*, 1990) all supported this qualitative prediction for heat-stimulated species, with strongest recruitment at high fire intensities. Since many legumes have heat-

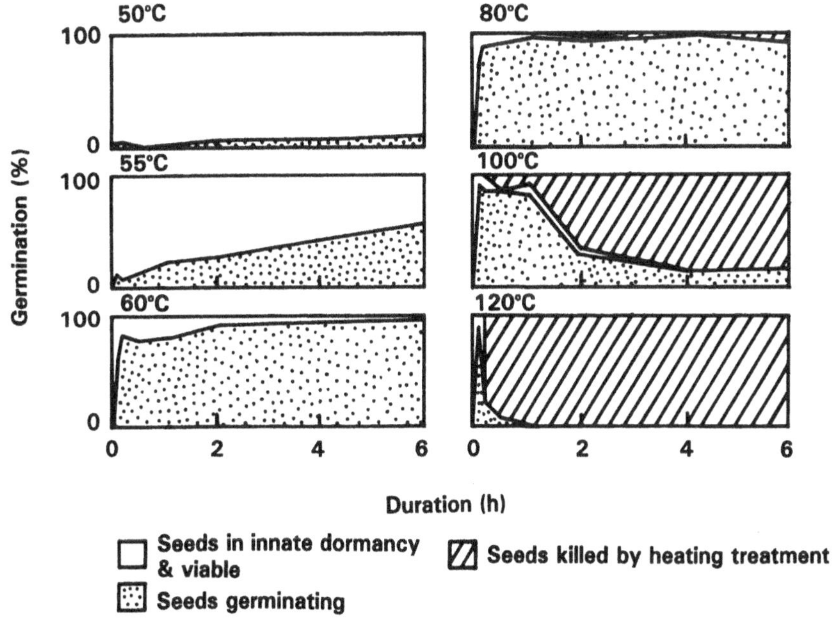

Figure 5.10 Response of seeds of *Acacia suaveolens*, an Australian shrub, to heating over a range of temperatures and durations (from Auld, 1986).

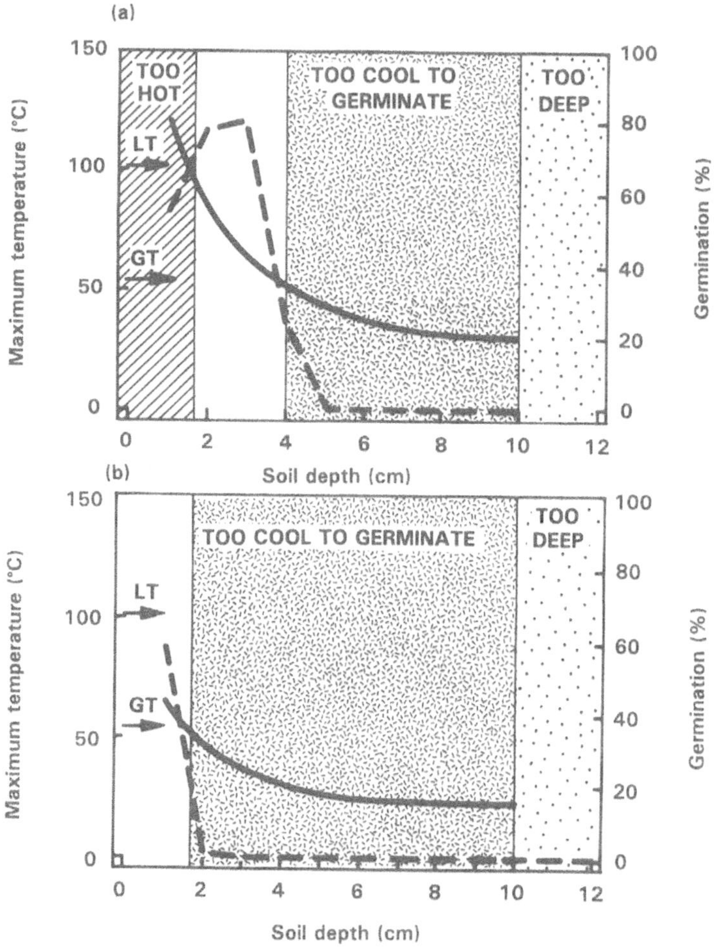

Figure 5.11 Predicted germination response of *Acacia suaveolens* in relation to soil depth and maximum temperatures reached after (a) a high- and (b) a low-intensity burn (from Auld, 1986). The temperature curves (solid lines) were derived from simulated hot and cool burns. The germination percentages (dashed lines) were predicted from germination data (see Figure 5.10). Germination fails if temperatures exceed a lethal threshold (LT: shaded as TOO HOT), temperatures fall below the minimum threshold needed to break dormancy (GT: TOO COOL) or if seeds are buried too deeply to emerge (TOO DEEP).

stimulated seeds, the widespread use of 'safe', low-intensity management fires could cause reduced legume recruitment, with far-reaching ecosystem consequences (Shea *et al.*, 1979).

Auld's model helps explain the diversity of recruitment responses seen after fire. First, seeds of different species may differ in the heat required to break dormancy. Laboratory studies have shown considerable species-specific variation in the temperature and duration required to stimulate germination. There seems to be far less variation in lethal temperature thresholds; few species tolerate temperatures above 150°C (Jeffery *et al.*, 1988; Auld and O'Connell, 1991; Keeley, 1991). Another source of variation is the depth from which seeds can emerge. Here seed size may be a useful general predictor of species tolerance of intense fires. Small seeds can only emerge from very shallow soil depths (see Figure 5.12 for an example from Cape fynbos) since size limits survival time on seed reserves alone (Molofsky and Augspurger, 1992; Westoby *et al.*, 1992). In high-intensity fires, lethal temperatures would penetrate below the maximum depth from which seedlings can emerge. Large-seeded species, as in Auld's studies, should escape intense fires by emerging from greater soil depths (Figure 5.12).

Auld's (1986) model is very useful in synthesizing the various factors influencing germination after fire. However, it should not be viewed as a tool for quantitative prediction of recruitment levels. This is because fire intensity itself is extremely patchy, especially in woody vegetation (Figure 5.13). The temperatures reached at various depths in the soil depend on

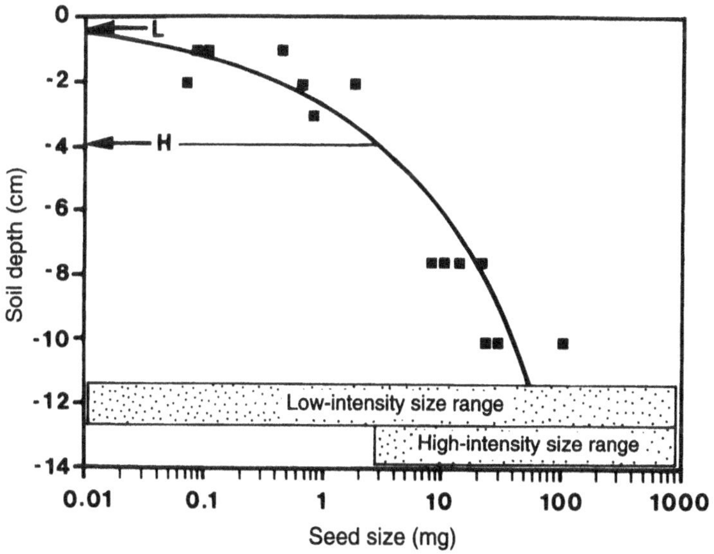

Figure 5.12 The relationship between seed size (mg) and maximum seedling emergence depth in sandy fynbos soils (Bond, Honig and Maze, unpublished). Squares indicate values for each of 14 species, representing 12 genera and 10 families. Seeds were germinated at 1, 2, 3, 5, 7.5 and 10 cm below the surface. Lethal temperatures would penetrate the soil more deeply during a high-intensity (H) than a low-intensity (L) burn. Small seeds surviving low-intensity burns would be able to germinate and their seedlings emerge (low-intensity size range). Small seeds surviving high-intensity burns may germinate but their seedlings would be too deeply buried to emerge (high-intensity size range).

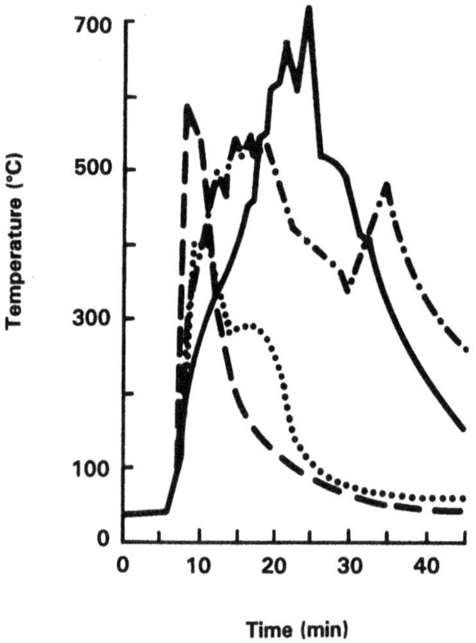

Figure 5.13 Local variation in fire temperature in a chaparral burn (DeBano *et al.*, 1979). Surface temperatures were recorded at four stations all situated within 90 cm of each other.

thermal conductivity, moisture status and soil structure, which are also patchy depending on soil patterns (Aston and Gill, 1976). The distribution of seeds in the soil is another source of variation. The value of Auld's model is its qualitative predictions. It suggests, for example, that the widespread shift of fire regimes to low-intensity, safe burns may have large impacts on community diversity, and we can begin to predict which kinds of species would be most affected.

5.9.2 Effects of phenology on recruitment

Recruitment of shrubs with soil-stored seeds varies not only with fire intensity but also with fire season (Le Maitre, 1988; Hodgkinson, 1991) but the mechanisms underlying seasonal variation are rather poorly understood. Two alternatives have been suggested: the temperature threshold argument discussed earlier (Auld, 1986) and the phenology of seedfall and subsequent rapid decline of soil seedbanks. Hodgkinson suggested that poor seedling recruitment after autumn and winter fires in an Australian eucalypt woodland was a consequence of poor heat penetration into wet soils, so that hard-seeded species failed to germinate. Alternatively, moist seeds may be less tolerant of heating than dry seeds, so that they are more easily killed in wet-season burns. In chaparral, imbibed seeds of both herb and shrub species were more easily killed by heating, possibly explaining their poor recruitment after wet-season fires (Parker and Kelly, 1989). Portlock *et al.* (1990) found that both factors may influence recruitment in a Western Australian legume. Heat penetrated to shallower depths in wet soils, effectively producing a 'low-intensity' germination response, but germination of wet seeds was also poorer than that of dry seeds.

It is not known how widely this 'wet seed' response accounts for seasonal variation in recruitment from soil seedbanks. An alternative argument is that recruitment differs after different fire seasons because of fluctuations in the seedbank (Keeley, 1977a; Pierce and Cowling, 1991). Variation in recruitment levels would then depend on the size of the seedbank immediately before the fire. We do not know of any studies that have shown a causal relationship between seasonal variation in seedbank size and post-burn recruitment levels. Soil seedbanks are difficult to study.

However, there is some support for this hypothesis in serotinous Proteaceae in South Africa and Australia. Two main causes for seasonal effects have been proposed for such species:

- the **pre-burn reserve** hypothesis argues that before fires, seedbank size fluctuates seasonally, from a peak after fruiting to a trough at flowering due to seed losses to predators, disease and decay;
- the **post-burn attrition** hypothesis argues that the seedbank decays from a maximum after fire-stimulated dispersal until germination in the appropriate season – the longer the delay between seed release and germination, the greater the seasonal effect (Bond, 1984) (Figure 5.14).

The first hypothesis is most appropriate for non-serotinous or weakly serotinous species. The second assumes that germination is delayed until favourable conditions for seedlings. In the summer dry climates of South African and Australian shrublands most germination is delayed to coincide with cool, wet winters (Bond, 1984; Cowling and Lamont, 1987). For both hypotheses, seasonal effects are indirect since they depend on the activities of seed predators or other agents of pre-emergence seed mortality. While this conveniently accounts for the considerable variance in fire-season effects (predation may vary in time and space), it adds to the difficulty of testing the hypotheses.

Cowling and Lamont (1987) tested the post-burn attrition hypothesis by comparing seedbank dynamics in four co-occurring *Banksia* species after a spring and autumn fire in Western Australian heath. All the species were serotinous and had no soil-stored seedbank. The fate of seeds was followed using exclosure experiments to measure predation effects. Seedlings appeared 2 months after an autumn burn when cool winter temperatures stimulated germination. However, after a spring burn, seeds remained dormant throughout the hot summer and only germinated in cooler winter conditions. The longer dormant period led to much higher losses to seed predators and poorer recruitment relative to autumn burns.

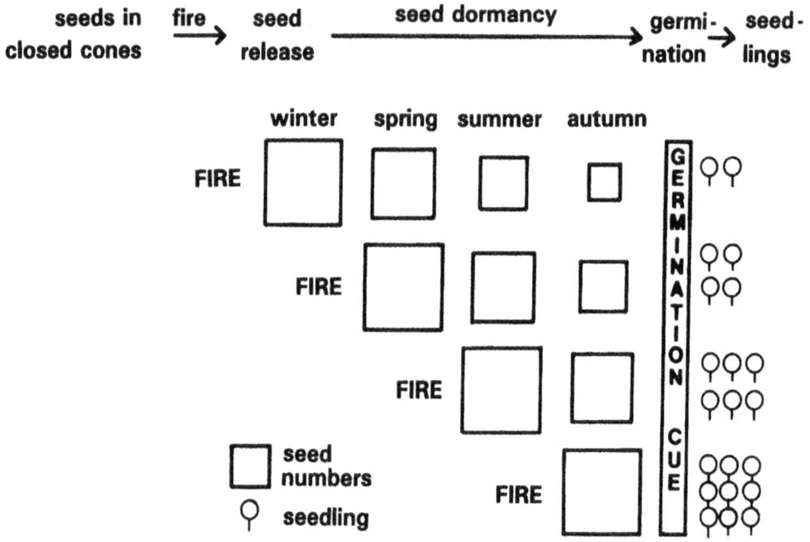

Figure 5.14 The post-burn attrition hypothesis for fire-season effects in serotinous species (Bond, 1984). Squares represent seed reserves, large after fires have stimulated seed release from serotinous cones but becoming progressively smaller due to attrition by predators, pathogens or insolation before they germinate in the cool rainy season.

Cowling and Lamont's study is generally consistent with the post-burn attrition hypothesis (Bond, 1984), first in the delay in post-burn germination and, second, in the proportionally greater seed losses where the delay was longest. Similar results of delayed germination leading to attrition of the seedbank have been reported in Cape fynbos and Australian heath (Bond, 1984; Midgley *et al.*, 1989; Bradstock and Bedward, 1992). Bond (1984) suggested that rodent seed predators were the main cause of seed losses in Cape fynbos. If post-burn predators were scarce, no season effect should occur. However, in at least some environments, seed losses were high in vertebrate exclosures, presumably due to summer scorching of seeds on the soil surface. This would produce a seasonal effect regardless of predators (Cowling and Lamont, 1987; Midgley, 1989; Midgley *et al.*, 1989).

5.10 PREDICTING EVENT-DEPENDENT EFFECTS

Some knowledge of event-dependent responses is important for managing natural communities. In many ecosystems, fire regimes have been greatly altered. Fires are burnt at lower intensities and in seasons less likely to produce runaway conditions. Habitat fragmentation no longer allows large fires to develop. The work reported in this chapter suggests that these changes will have substantial effects on some communities. It would be very convenient to identify which species would be most likely to be affected.

The studies reported in this chapter show the difficulty of prediction. Species responses are highly individualistic. Even within species, responses vary among populations and along moisture gradients, for example in grass response to fire season (Glenn-Lewin *et al.*, 1990). Responses differ depending on the disturbance parameter – the effects of fire season depend on phenology, the effects of fire intensity on the location of meristems. Responses are sometimes indirect and depend on interactions with other

Table 5.8 Functional types for event-dependent effects. The relative response of each type to fire intensity and season is indicated by the number of crosses, or 0 where there is no effect. Question marks indicate little empirical evidence. Intensity effects are produced by variation in the heat released by a burn. Season effects are those relating to the phenology of the plant or its carbon and nutrient reserves. Season effects may also be caused by poor germination of heated, imbibed seeds – 'wet heat'. Where season effects are thought to be due to heat yield of a burn, they are listed only under intensity. Note that all seedlings can be killed by intense fires

Functional type	*Intensity*	*Season*	*Mortality mechanism*	
			Intensity	*Season*
Fire survival mode				
Above-ground buds				
(a) Existing shoot apices	XXX	X	Killed by crown fires	Resources for sprouting
(b) Epicormic buds	0	XX	–	Resources for sprouting
Below-ground buds				
(a) Superficial buds	X (X)	XX	Buds killed	Resources for sprouting
(b) Deep buds (e.g. root suckers)	0	XX	–	Resources for sprouting
Recruitment				
Soil seedbank				
Germination not fire-stimulated	0	0		
Germination fire-stimulated				
(a) Long-lived soil seedbank				
(i) Heat-stimulated	XXX	X?	Lethal heat, germination stimulus	'Wet heat' germination stimulus
(ii) Other stimulus small seeded	XXX?	X?	Lethal heat, shallow emergence depth	'Wet heat'
large seeded	X	X?	Lethal heat	'Wet heat'
(b) Short-lived soil seedbank	XX	XX?	Lethal heat	Reproductive phenology
Canopy seedbank	0 (X)	XXX	–	Reproductive phenology

species (Chapter 7). Finally, there is considerable intrinsic spatial and temporal variability within and between fires and the populations they affect (e.g. Figure 5.13 for variation in fire temperatures in a chaparral burn). Fire is a messy natural phenomenon and the spatial and temporal scales of prediction are limited.

Nevertheless, studies of possible mechanisms underlying variation in survival and recruitment do suggest a framework for predicting fire effects. In Table 5.8 we list a tentative set of vital attributes that provide qualitative predictions of likely fire-season or fire-intensity responses. For fire-intensity response, the key attribute is the location of bud tissues, or seeds, in relation to heat sources. For season of burn effects, the key attributes relate to the phenology of the plant or the (still poorly researched) germination response of wet or dry seeds. Season of burn effects are listed in the table only where they appear to act independently from intensity effects.

Table 5.8 complements Table 4.4 for fire-interval effects. Both provide a qualitative set of species response rules for fire-prone vegetation.

5.11 EVENT-DEPENDENT EFFECTS IN DIFFERENT BIOMES

The relative importance of fire season or intensity in determining vegetation change differs among major plant biomes. Where fuels are relatively uniform, fire intensity varies too little to be of much significance. Grassland fire intensities are least variable (range from 1000 to 3000 $kW\,m^{-1}$), followed by savanna fires (500–10 000 $kW\,m^{-1}$). Fire intensity is much more variable in shrublands (ranging from less than 500 to greater than 20 000 $kW\,m^{-1}$ for fynbos shrublands) and extremely variable in conifer forests (500–100 000 $kW\,m^{-1}$ or more in crown fires).

Biomes differ too in the variability in fire season, depending on the duration of hot, dry periods. Grass fuels dry quickly permitting a wide variation in fire season. Australian and South African shrublands, with a graminoid understorey, also burn in most seasons (Brown *et al.*, 1991). Chaparral appears to have a much narrower fire season and seldom burns in the wet season. Fire season is also of short duration in many temperate and boreal forests, especially those with regular winter snow.

These fuel and climate constraints on the fire regime affect the relative importance of fire season and fire intensity in different biomes. Fire season has a marked regulatory effect in grasslands such as tallgrass prairies and tall grassveld, at least some savannas (Trapnell, 1959), and Australian and South African sclerophyll shrublands. Seasonal effects are also important in controlling understorey shrubs in some conifer forests. Fire intensity, in contrast, is very important in determining survival of canopy trees in conifer forests. Different intensities can drastically alter age and species composition of forests burnt at similar frequencies (Heinselman, 1981). Fire intensity is also important in shrublands, affecting both survival and recruitment.

Plate 1 Mass flowering of *Watsonia borbonica* stimulated by an autumn burn in South African fynbos. (G.G. Forsyth)

Plate 2 Serotinous cones on a *Banksia attenuata* shrub. The swollen woody follicles insulate the seeds during a fire and release them into the burnt vegetation. The demography of *Banksias* and other serotinous shrubs has been extensively studied, partly because seedbanks are easily estimated. (W.D. Stock)

Plate 3 Post-burn succession produces age-related changes in vegetation structure which influence landscape pattern. The foreground of this landscape in Glacier National Park, U.S.A., has experienced a burn 15 years previously, the middle ground is recovering from a burn 72 years previously and the background has not been burnt in recorded history. (B. van Wilgen)

Plate 4 A fire mosaic in boreal North American conifer forests. Consistent differences in fire regimes, such as lower intensity burns along lake margins, can cause species turnover as a result of different fire tolerance. (B.J. Stocks)

Plate 5 Non-flammable forest persisting along a water course in a fire-prone fynbos landscape in South Africa. Forest species can spread to the adjacent slopes after long periods of fire exclusion. (B. van Wilgen)

Plate 6 A fire-dependent landscape in the Drakensberg, South Africa. Grassland fires restrict the spread of forests from riparian or rocky refugia. (B. van Wilgen)

Plate 7 Low intensity, surface fires in the boreal conifer forests of Siberia. (J. Goldammer)

Plate 8 High intensity crown fires in the boreal conifer forests of Siberia. The low canopy height of these trees provides a 'ladder' for the fire to reach the forest canopy. (J. Goldammer)

In savannas, fire intensity influences topkill of trees and the grass/tree balance, but the effect is confounded with fire season. Variation in intensity is slight in grasslands and seems relatively less important than in other biomes.

5.12 CONCLUSIONS

In Chapter 4, we began by contrasting interval-dependent, density-dependent and event-dependent causes of population fluctuations in fire-prone environments. It is too early to judge the importance of density-dependent effects. However, the large number of studies, using a variety of methods, in many parts of the world and based on agricultural, forestry and ecological literature, leave little doubt that event-dependent effects are important in regulating plant numbers, and that these effects often occur independently of disturbance frequency. Both must be considered for any complete understanding of the importance of fire in shaping vegetation.

We suggest that fire frequency determines population changes at very short and very long intervals, but that fire intensity and fire season are more important at intermediate frequencies. There is growing understanding of the mechanisms underlying event-dependent fire responses. However, the stochastic nature of fire ultimately limits the success of attempts at prediction at these intermediate fire frequencies.

The magnitude of event-dependent effects is important from the theoretical perspective of whether fire-prone communities are equilibrial or non-equilibrial systems (de Angelis and Waterhouse, 1987). Ecological systems can be ranked on a scale of decreasing stability, caused either by disruption from internal feedbacks, such as strong density dependence (Chapter 4), or from stochasticity (de Angelis and Waterhouse, 1987). Fire does not necessarily generate instability. System composition may remain highly stable and predictable within a specified fire regime. For example, some tall grasslands, mesic savannas and French garrigue shrublands change little after successive fires burnt at the same frequency and season. Burning leads to only transient effects on system composition (Trabaud, 1991).

Stochastic effects are greatest where fire intensity is most variable, usually because of patchy fuels or highly variable rates of fuel accumulation. Fire-dependent stochasticity is common in flammable woodlands and forests, some shrublands and the more arid savannas (Cowling, 1987). Here fires burnt at similar intervals and seasons can have very different effects between fires and between patches in the same fire. Climate variation adds further unpredictability. Spatial and temporal heterogeneity further confounds quantitative predictions of population change. The recent shift of emphasis to models that explore population persistence, rather than size, seems very useful. Even if a species fails to survive a burn in one patch after a fire, it may persist in another patch with different fire behaviour. The contingencies of the fire event and the spatial scale across which populations can persist become central concerns (de Angelis and Waterhouse, 1987; Chapter 7). Determining just where different fire-prone systems lie on the gradient from stable equilibrium to fire-driven stochasticity, and exploring the implications for theoretical and applied questions in ecology, are rich challenges for the future.

6 Fire and the evolutionary ecology of plants

6.1 FIRE AND THE EVOLUTION OF PLANT TRAITS

Evolutionary ecology attempts to understand the functional significance of a trait by searching for pattern and attempting to explain the underlying cause. Fire is only one of many evolutionary forces that shape plants. Are there any plant traits peculiarly associated with fire? How does fire shape life history, morphology, and reproduction? Do differing fire regimes favour one set of traits over another? Until recently, evolutionary questions were neglected in fire research despite frequent reference to some traits as fire 'adaptations'. But some so-called fire adaptations, such as thick bark or the capacity to sprout, occur in many taxa in places where fire is alien, such as tropical rainforest (Gill, 1981b; Uhl and Kaufman, 1990). New studies are more cautious. Traits are compared only in closely related taxa and phylogenetic methods are beginning to be used to trace the history of trait evolution (Midgley, 1987; Linder, 1995). Even with tighter phylogenetic control, it is clear that distinct patterns of recurring features – adaptations – occur in fire-prone regions in unrelated taxa and widely separated geographic regions. These form the subject of this chapter.

Among the most distinct fire adaptations are the suite of characters associated with fire life-histories – the scheduling of reproduction and mortality. Population biologists have long been interested in the variety of plant life-histories and their distribution patterns in different environments (Gadgil and Solbrig, 1972; Werner and Platt, 1976; Willson, 1983; Loehle, 1988; Tilman, 1988). Why does tree life span vary from a few years to millennia? When should plants first reproduce? For how many reproductive episodes? With how many seeds and of what size? When should they germinate, now or later? The same questions are relevant in studies of fire life-histories but with different emphases and some interesting differences. For example, the evolution of monocarpy from polycarpy becomes, for fire-prone environments, the evolution of non-sprouting from sprouting ancestors.

Fire ecologists have also studied the evolution of flammability itself. Just as many ecologists have studied plant secondary chemicals and inferred fitness advantage in relation to herbivory, so too a much smaller number of fire ecologists have studied secondary chemicals in relation to flammability and also inferred evolutionary advantage to being burnt. There is good evidence that some plants and plant communities burn more easily than others and that their persistence is favoured by fire (Chapter 2). Has flammability evolved because it is advantageous to be burnt? The argument is at least as controversial as the similar debate on whether grasses might have evolved to favour being eaten (McNaughton, 1984; Belsky, 1986).

Plants in fire-prone environments offer a largely untapped resource for addressing evolutionary questions relating, for example, to fitness consequences of different plant architectures, breeding systems, pollinators, or dispersal modes. Fitness components are quite easily measured in fire-recruiting plants with their discrete generations. In this chapter, however, we address only those evolutionary questions involving constraints and trade-offs **unique** to fire ecology.

6.2 SPROUTERS VERSUS NON-SPROUTERS

One of the central questions in life-history theory is what is the optimum frequency of reproduction. Most plants reproduce more than once in their lives, they are **iteroparous**. However, some, including perennials such as bamboos, flower only once and die. They are **semelparous**. The analogous situation in fire environments is the contrasting sprouting and non-sprouting habit of fire-recruiting species. For both types, flowering may occur annually but recruitment is triggered by fire. Sprouters survive repeated fires and therefore experience repeated regeneration episodes – non-sprouters are killed by fire and only experience one. There has been considerable interest in the circumstances favouring the one or the other survival type, especially in shrubland vegetation where both types are common. Sprouters and non-sprouters often share similar size, general morphology, physiology and habitat, yet non-sprouters frequently dominate communities. This seems counterintuitive given the apparent advantage of sprouters in (a) not relinquishing occupied space to competitors after fire and (b) their insurance against recruitment failures.

Sprouting in woody plants is widely assumed to be the ancestral condition (Wells, 1969) and failure to sprout, the derived. However, there are exceptions (Midgley, 1987) and both life histories may be selectively advantageous under particular circumstances. Here we explore these advantages in the context of classic life-history theory (Stearns, 1977; Willson, 1983).

6.2.1 Allocation patterns and the cost of sprouting

At the heart of life-history theory is the principle of allocation. Every organism allocates its resources to various essential activities such as maintenance, growth and reproduction. Natural selection, according to the principle of allocation, should favour allocation patterns that maximize fitness of the individual. How do allocation patterns differ in sprouters and non-sprouters and how do they influence fitness? In fire-prone environments allocation to fire survival, rather than reproduction, is the major cost. To survive fires, sprouters must allocate resources to protective bark or to storage reserves in roots or shoots to replace damaged tissue. Allocation to root reserves must occur early in the growth of sprouter seedlings if they are to survive their first fire. Non-sprouters shed these survival costs and can therefore gain reproductive and growth rate advantages.

Fitness advantages for non-sprouters may be expressed in a number of ways including greater fecundity, longer-lived seedbanks, differences in juvenile survival or high sprouter mortality (Box 6.1). Most studies have examined only one or two of the possible variables. It is hardly surprising, then, that a variety of hypotheses have emerged explaining sprouter–non-sprouter ecology tailored for different sets of species and that the supposed fitness advantages of non-sprouters have not always been demonstrated.

Pate and co-workers have made particularly comprehensive studies of allocation and growth differences between numerous species of sprouters and non-sprouters from south-west Australian heathlands. In comparisons of juveniles (2–4 years old) from congeneric pairs of sprouting and non-sprouting species, Pate *et al.* (1990) found that sprouters had:

- slower shoot growth;
- lignotubers but less extensive root systems;
- delayed flowering (from the fourth to the seventh year of growth, compared to only the second or third year in non-sprouters).

For a large sample of species, sprouters had mean shoot:root ratios of 1.2 (14 spp.) compared with 5.0 for non-sprouters (18 spp.) and much lower growth rates (total dry weight of 3.9 ± 0.9 g versus 10.7 ± 2.2 g for comparably aged non-sprouters). Sprouters had much higher concentrations of starch in the roots – typically four times that of non-sprouters. The Myrtaceae were an interesting exception with no significant difference in starch reserves in roots of the two types. Energy appeared to be the main currency of allocation since there were no differences in concentrations of nitrogen, phosphorus, potassium, calcium and magnesium in either roots or shoots of the two life-history types.

Non-sprouters in these Australian shrubs allocated about four times more assimilates to above-ground parts than sprouters (Figure 6.1). Pate *et al.* (1990) comment that since 'storage of starch in shoot or root has no immediate relevance to furthering growth, a root-dominated starch-storing resprouter is clearly disadvantaged in growth rate when making such a small investment of its resources into a photosynthesizing shoot'.

Very similar allocation patterns have been found in a large sample of Restionaceae, a family of reed-like plants (Pate *et al.*, 1991). Here too, comparison of juvenile sprouting and non-sprouting species indicated distinct differences in resource allocation in the two groups. Non-sprouting

Box 6.1 A simple model for conditions under which a non-sprouter would evolve

The importance of adult and juvenile survival for the evolution of fire life-histories can be illustrated using a simple deterministic model following Charnov and Shaffer's (1973) analysis of semelparity and iteroparity. If the fire interval is assumed constant, and longer than the age to maturity, the growth of a sprouting population, N, from one fire to just before the next can be expressed as,

$$N_s = Ap + c_s AB_s \qquad (6.1)$$

where A is the number of adults before the last fire, p is the probability of a mature plant surviving through this fire until the next, B_s is the seed crop when burnt, and c_s is the proportion surviving to germinate and grow to maturity. A mutant non-sprouting genotype, N_n, has $p = 0$ and the growth rate of its progeny can be expressed as

$$N_n = c_n AB_n \qquad (6.2)$$

where B_n is the seed crop of the non-sprouter and c_n is the proportion surviving to maturity. The non-sprouting genotype would have a fitness advantage if $N_n > N_s$ or, substituting from equations 6.1 and 6.2,

$$c_n AB_n > Ap + c_s AB_s.$$

Dividing both sides of the equation by $c_n A$,

$$B_n > p/c_n + (c_s B_s)/c_n.$$

This means that for a non-sprouter to be favoured over a sprouter, it must produce sufficient seeds, B_n, to exceed the right-hand expression. The implications of the equation for life-history evolution are illustrated using an example from chaparral comparing a species pair rather than a 'mutant' non-sprouter (ex Keeley, 1977a, see Table 6.2). In this simplified example, with $p = 1$, $c_s = 0.0045$ (post-burn seedlings/pre-burn seedbank), $c_n = 0.0054$ and $B_s = 83.5$, the non-sprouter would need a seedbank of 254 seeds to exceed the sprouter, compared to an actual value of 262 for *C. greggii*. The amount of seeds needed by a non-sprouter to beat a sprouter under different values of p, c_n and c_s are shown in the figure opposite. The cost of not sprouting is very sensitive to juvenile survival rates (c_n), less sensitive to adult sprouter mortality rates and least sensitive to sprouter juvenile mortality (c_s).

The gain in juvenile survival c_n, or fecundity, B_n, of the non-sprouter is presumed to be derived from resources saved by not sprouting. However, B_n can also be modified through changes in seed longevity in the seedbank.

(continued)

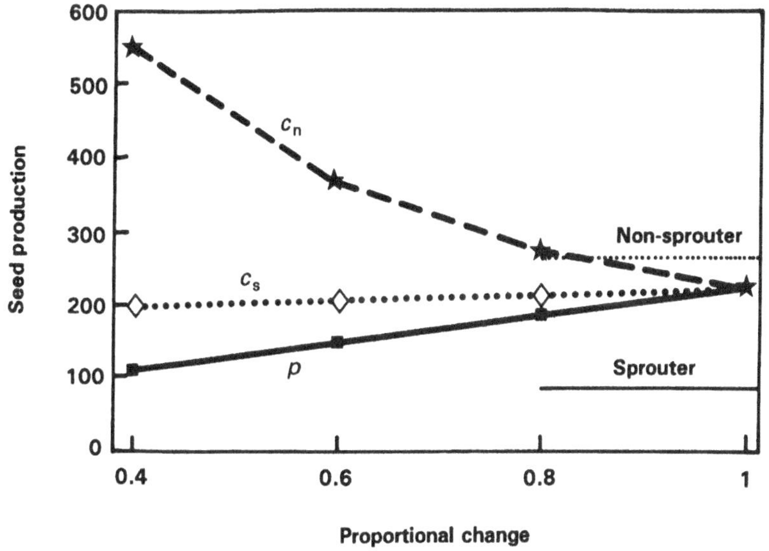

The sensitivity of a simple model for the evolution of a non-sprouter from a sprouter. They *y* axis indicates B_n, the seed production of the non-sprouter needed to beat its sprouting ancestor. Parameter values are taken from *Ceanothus spinosus*, a sprouter, and *C. greggii*, a co-occurring non-sprouter (Table 6.2; Keeley, 1977) where *p* is the probability of adults surviving from one fire to the next, c_s is the probability of seeds surviving a fire and producing seedlings and c_n is the same for non-sprouters. The lines marked sprouter and non-sprouter indicate pre-burn seedbank sizes for the two species. Note that seed production needed by the non-sprouter to beat the sprouter increases markedly with any reduction in its seedling survival rates.

species, in congeneric comparisons, were more likely to have taller culms, less deeply buried rhizomes and these of narrower diameter than sprouters. Culm to rhizome biomass ratios were, as in the shrubby species, higher in the non-sprouters (mean of 12.3 versus 2.9 for sprouters). Also, as in the shrub sample, starch concentrations in below-ground storage organs were significantly higher in sprouters (Figure 6.2).

Both these studies provide clear evidence for trade-offs between allocation to roots for future fire survival in sprouters and to shoots for immediate seedling growth in non-sprouters. The patterns are remarkably convergent across unrelated families.

6.2.2 The fitness consequences of allocation trade-offs

Pate's studies are among the first to demonstrate directly the costs of sprouting in fire-prone species (but see Keeley and Keeley, 1977 for a chaparral study). Many more studies have assumed these costs and ex-

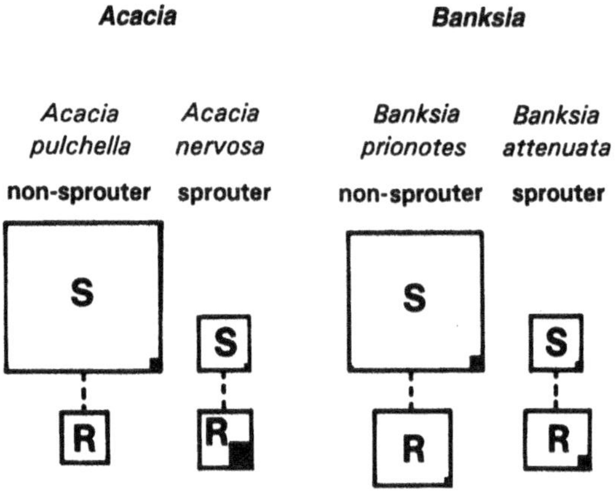

Figure 6.1 A comparison of partitioning of dry matter and starch (shaded) between shoot (S) and root (R) of sprouting and non-sprouting species pairs of Western Australian shrubs. The size of each box indicates relative allocation. Both species pairs co-occur. (Data from Pate *et al.*, 1990.)

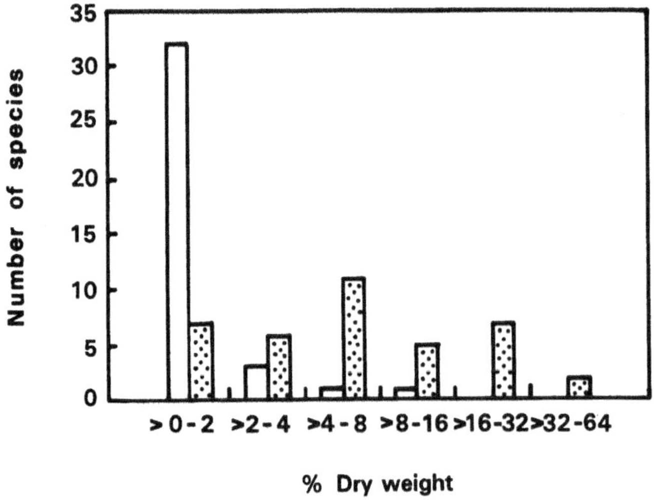

Figure 6.2 Assimilate storage patterns contrasted in sprouting (shaded) and non-sprouting (unshaded) Restionaceae (from Pate *et al.*, 1991). Starch and sugar were measured in summer-dormant rhizomes and are expressed as percentage dry weight. Values for 38 sprouters and 37 non-sprouters are shown.

plored the fitness consequences, usually by comparing congeneric species pairs. These can be manifested as differences in juvenile survival, the scheduling of reproduction, the number of flowers, fruits and seeds, the longevity of seedbanks and the life span of plants.

Juvenile mortality should generally be higher in sprouters than non-sprouters since growth is stunted by the cost of producing storage tissue for

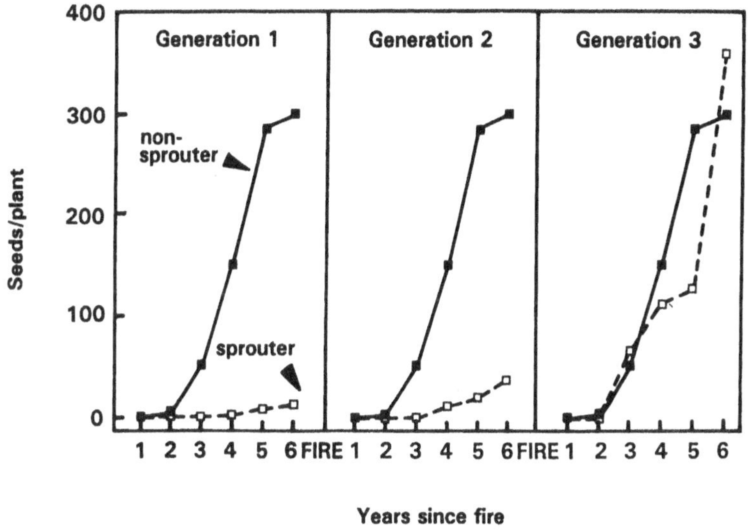

Figure 6.3 Comparison of reproduction in a sprouting (*Bossiaea ornata*) and non-sprouting (*B. aquifolium*) pair of co-occurring leguminous shrub species from Western Australia. The graph contrasts seed production of three successive generations of non-sprouters (closed squares) with a single cohort of sprouters (open squares) recovering after three successive fires. Fire interval was approximately 6 years. (Data from Hansen *et al.*, 1991b.)

fire survival. In a chaparral study, only 18% of the seedlings of *Ceanothus spinosa*, a sprouter, survived the first 2 years after a burn, compared to 42% in the co-occurring *C. megacarpus*, a non-sprouter (Thomas and Davis, 1989). Similar trends of higher mortality in sprouter seedlings were found in simulated droughts applied to two *Banksia* species (Zammitt and Westoby, 1987). However, in a field experiment, in a particularly wet year, there were no mortality differences in the same species pair. Le Maitre (1992) found no apparent differences in juvenile mortality in a field comparison of a pair of sprouting and non-sprouting fynbos *Protea* species.

Differences in reproductive scheduling have been studied in two co-occurring leguminous shrubs in south-west Australia, one, *Bossiaea aquifolium*, a non-sprouter and the other, *B. ornata*, a sprouter (Hansen *et al.*, 1991b). The non-sprouter began flowering much earlier and had much higher fecundity than the sprouter (Figure 6.3). Indeed, the sprouter only caught up after surviving three fires. For this species pair, sprouting appears to have a profound reproductive cost relative to the precocious non-sprouter. Similar results have been reported in comparisons of species pairs in the Proteaceae (Lamont, 1985; Le Maitre, 1992) and in pines (Clark, 1991). The implication is that a mutant non-sprouter would rapidly replace its sprouting ancestor since it would leave more seedlings after each burn. Just why sprouters still exist given the costs involved is discussed in section 6.2.3.

Table 6.1 Seed numbers of sprouting and non-sprouting species in Western Australian shrublands. Species pairs with the same number shared the same site. Asterisked ratios were significantly different (t test $P < 0.05$). (Data ex Bellairs and Bell, 1990)

Species		*Seed number*		*A/B*
Non-sprouters	*Sprouters*	*Non-sprouter (A)*	*Sprouter (B)*	
1. *Allocasuarina campestris*	*A. acutivalvis*	11667	1991	5.8*
1. *Melaleuca scabra*	*M. tuberculata*	3634	1020	3.6
Melaleuca pungens	*M. tuberculata*	3355	1020	3.3
2. *Melaleuca scabra*	*M. tuberculata*	2592	291	8.9*
3. *Banksia hookeriana*	*B. attenuata*	347	23	15.1*
Banksia hookeriana	*B. grossa*	347	31	11.2*
Banksia hookeriana	*B. menziesii*	347	52	6.7*
4. *Banksia hookeriana*	*B. attenuata*	101	12	8.4*
Banksia hookeriana	*B. menziesii*	101	21	4.8*
4. *Hakea obliqua*	*H. corymbosa*	13	4	3.3

Reproductive effort, the number of flowers, fruits and seeds per unit biomass, should generally be greater in non-sprouters. Many studies have compared reproduction in species pairs, at a single post-fire age, without factoring in relative size. Even with these rather crude comparisons, non-sprouters produced more seeds and had larger seedbanks than congeneric sprouters (Specht *et al.*, 1958; Carpenter and Recher, 1979; Bellairs and Bell, 1990; Le Maitre, 1992; see Table 6.1). In some cases, however, the reverse was true and sprouting plants had more seeds (Keeley, 1977a; Lamont, 1985; Table 6.2).

Another way of looking at reproductive trade-offs between sprouters and non-sprouters is to compare the degree of sprouting with seedling production. Keeley and Zedler (1978) described chaparral shrub life histories as a continuum from obligate sprouting to weak facultative sprouting to non-sprouting. They found evidence for the expected trade-off between levels of sprouter mortality and number of seedlings produced after a fire. Data from a fynbos fire indicated a similar trade-off over a range of unrelated families (Le Maitre *et al.*, 1992; Figure 6.4). This trade-off between persistence (of sprouting individuals) and recruitment may be quite general for fire and other disturbances.

Seedling mortality, reproductive scheduling and fecundity are just three

Table 6.2 Comparison of seed production, seed storage in the soil and seedlings produced after fire in sprouting (S) and non-sprouting (NS) species of *Ceanothus* and *Arctostaphylos* species pairs (ex Keeley, 1977a). Note the large inter-annual variation in seed production

Species	*Annual seed production* m^{-2}					*Soil seedbank* m^{-2}	*Post-burn seedlings* m^{-2}
	1972	*1973*	*1974*	*1975*	*1976*	*(1972)*	
Ceanothus greggii (NS)	–	560	8350	0	–	262	1.44
Ceanothus leucodermis (S)	–	145	1280	2	–	83	0.21
Arctostaphylos glauca (NS)	10	4	853	0	130	187	0.85
Arctostaphylos glandulosa (S)	144	0	2200	0	104	288	0.15

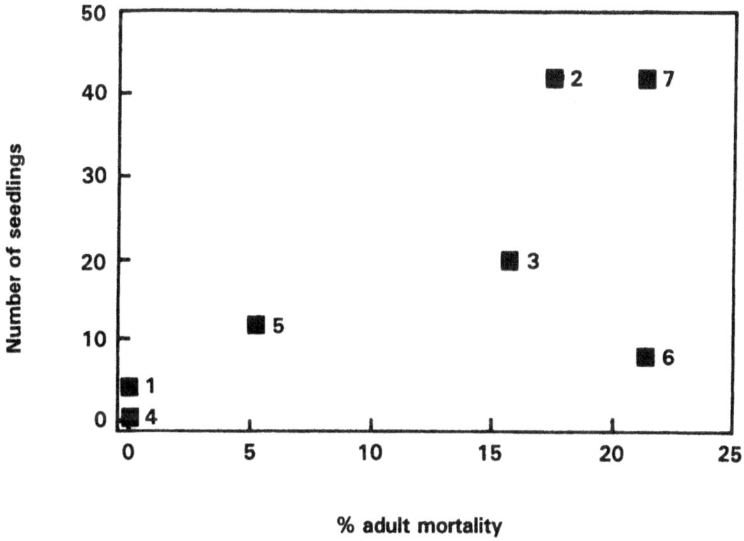

Figure 6.4 The relationship between fire-induced mortality (%) and seedling recruitment in sprouting fynbos shrubs (from Le Maitre *et al.*, 1992). Seedlings were censused 2 years after an autumn fire. Species are: 1, *Widdringtonia nodiflora*; 2, *Penaea mucronata*; 3, *Mimetes cucullatus*; 4, *Aulax pallacea*; 5, *Nebelia paleacea*; 6, *Erica coccinea*; and 7, *Leucadendron salignum.*

of the several potential demographic trade-offs between seeders and sprouters. Most authors have found clear support for the predicted trade-offs (e.g. Hansen *et al.*, 1991b) but there are exceptions (e.g. Lamont, 1985). Allocation trade-offs may occur in different ways in different species, depending on their morphology, reproductive biology and developmental patterns. This makes it hard to falsify the notion of demographic trade-offs since ultimately only a complete analysis of all demographic parameters influencing fitness is good enough. As yet no such studies exist, though they may be easier in fire-prone shrublands than in just about all other perennial plants.

Another difficulty with studies of trade-offs is the arbitrary choice of species pairs. Ideally these should be sister species, or preferably intraspecific variants. Without controlling for phylogeny, it becomes difficult to attribute life-history differences to the sprouting/non-sprouting dichotomy rather than to other species attributes.

6.2.3 Selective regimes and the evolution of alternative fire-survival types

Let us recap briefly the attributes of fire-recruiting sprouters and non-sprouters. Sprouters, once they have appropriated space, occupy it over a number of successive fire cycles. With their large established root systems, they grow rapidly after fires and are almost immune to competitive suppression by non-sprouters in the early post-burn years. In addition, sprouting plants soon recover their reproductive capacity. In contrast, seedlings suffer all the disadvantages of small size, and cannot displace established plants. They face, too, a longer time lag before flowering can begin. However, non-sprouters are at an advantage when competing for open space. By allocating resources to growth and reproduction, mature plants produce more seeds for colonizing gaps over longer fire intervals. More importantly, their seedlings have faster growth rates and outgrow sprouter seedlings.

Under what circumstances would we expect one or other of the two life histories to be favoured? Two hypotheses have been suggested, both linking non-sprouter success to the availability of open space. The fire-frequency hypothesis (Keeley, 1977b; Keeley and Zedler, 1978; Keeley, 1986) relates the two life histories to fire regime. As a population of sprouters grows after a fire, thinning should occur resulting in increasing distances between individuals. After a short fire interval, survivors will still be closely packed with little space for seedlings. As intervals between fires increase, sprouters should become more widely spaced, both as a result of thinning and of increased mortality from the more intense fires. The result, after a fire, will be more open space for seedlings. Longer fire intervals should thus favour non-sprouters (see Figure 6.5). Distribution patterns of chaparral shrubs in California appear to support these predictions. Sprouter densities are higher in the Sierra Nevadas and at high elevations where lightning is common and fires frequent. Non-sprouters predominate in

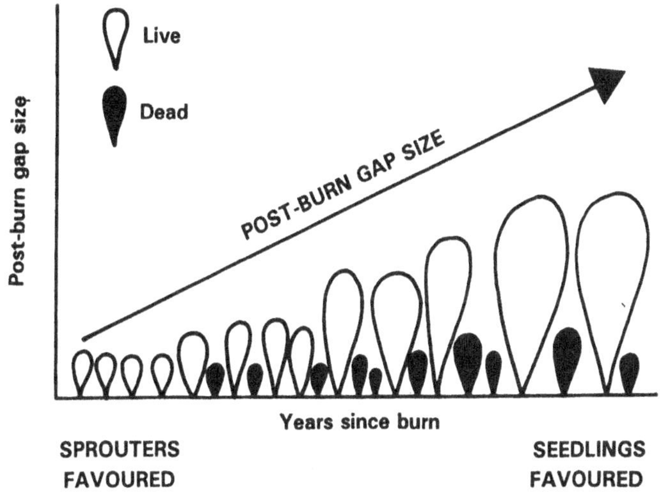

Figure 6.5 The fire frequency hypothesis for sprouter/non-sprouter dominance (from Keeley, 1977a; Keeley and Zedler, 1978). Non-sprouter seedlings are assumed better competitors for unoccupied space than sprouter seedlings. Post-burn gap size increases with longer intervals between fires because of interfire thinning and increased shrub mortality after high-intensity burns.

southern California and at low elevations where lightning is rare and fire intervals longer (Keeley, 1977b).

A similar argument underlies the gap-availability hypothesis, which also relates recruitment opportunities to the availability of gaps, but in relation to resource gradients. Specht (1981) noted that 'few seedlings normally survive, other than in "gaps" where, for some reason, the parent community has degenerated'. He found that the proportion of above-ground biomass made up of sprouters was linearly related to the cover of the upper stratum in heathland communities. Sprouter dominance increased along a moisture gradient, presumably as plant cover thickened up. Recent studies in fynbos shrublands found a very similar correlation between sprouter biomass and a soil moisture gradient (Smith *et al.*, 1992).

Both the fire frequency and gap availability hypotheses assume that space limits recruitment. This assumption has yet to be tested in post-burn conditions. In an apparent reversal of cause and effect, some authors have argued that the changing abundance of sprouters and seeders across fire regimes is a **result** of their life-history differences. Kruger and Bigalke (1984) noted that non-sprouters have a vulnerable pre-reproductive period and, at least for fynbos Proteaceae, a shorter life span than sprouters. They will therefore be eliminated not only by very short fire intervals but also by very long ones exceeding the life span of plants and their seedbanks. Similar arguments were used by Hilbert (1987) who modelled fitness of sprouters and non-sprouters based on chaparral shrub life-histories. His model, based only on demography and with no competition for space, also predicted

Table 6.3 Number of sprouting and non-sprouting taxa in several shrubland genera (% of genus in parentheses)

Locality	*Genus*	*Sprouters*	*Non-sprouters*
California[a], chaparral	*Ceanothus*	12 (20.7)	46 (79.3)
	Arctostaphylos	16 (21.3)	59 (78.7)
Australia[b]	*Banksia*	(49.0)	(51.0)
	Hakea	(52.0)	(48.0)
South Africa[c], fynbos	*Leucadendron*	9 (11.4)	70 (88.6)
	Leucospermum	12 (26.1)	34 (73.9)
	Protea	20 (28.6)	50 (71.4)

[a] Wells (1969).
[b] Lamont *et al.* (1985).
[c] Williams (1972); Rebelo (personal communication, 1994).

optimal fire frequencies at intermediate length for non-sprouters. The model is exemplary in indicating the complexity and the number of parameters needed to predict fitness differences between fire life-histories. Neither Hilbert, nor Kruger and Bigalke explain why non-sprouting life-histories should have evolved in the first place, nor why they often dominate their communities at their preferred fire frequency.

The least-studied hypothesis is that suggested by Wells (1969). He argued that non-sprouters would have a greater number of sexual generations, resulting in more rapid natural selection and closer genetic tracking of changing environmental conditions. In a way, the two life histories resemble sexual and asexual breeding systems: non-sprouters test new genetic combinations after each fire while sprouters preserve the same combinations for many fires. Unfortunately, the genetic implications of fire life-histories have not been studied further. This must be one of the few life-history combinations for which no population genetic analyses or allozyme studies are available (see, for example, Hamrick and Godt, 1990). Speciation may also be influenced by fire life-history. Wells (1969) noted a marked preponderance of non-sprouting species in Californian shrub genera with both types represented, though the pattern is by no means universal (Table 6.3). This is clearly an area deserving further investigation.

6.3 LIFE-HISTORY EVOLUTION

6.3.1 Age to maturation and life span

As noted in Chapter 4, many woody plants in fire-prone systems senesce and die after long intervals without fire. What determines the plant life spans, and how might they be related to the prevailing fire regime? Answers to this question also implicate different patterns of resource allocation.

Because resources are finite, allocation to reproduction or survival are to some extent alternative options. The central assumption of life-history studies is that allocation of resources to reproduction carries a cost of reduced growth and survival, and allocation to survival results in reduced reproduction. Loehle (1988) has explored these trade-offs extensively in plants. He argued that, in order to live longer, trees must allocate resources to structural support to withstand wind, bark to withstand fire and other damage and structural and chemical defences to withstand pathogens that attack wood. However, because of the trade-offs between reproduction and survival, longevity should be inversely related to the onset of maturity. Loehle analysed a large set of data on life span, maturation time, and energy investment in wood for North American angiosperm and gymnosperm trees. He found, as predicted, that life span was positively correlated with energy investment in wood and negatively correlated with maturation age. Plants pay for early reproduction by dying sooner and for longevity by reproducing later. His analysis strengthens and extends earlier studies of plant life-history correlations (Harper and White, 1974).

In fire-prone environments, life spans can be terminated abruptly by canopy fires. In fire-recruiting species (Chapter 4) recruitment is mostly restricted to the brief post-fire period. Plants that gamble all their resources on precocious reproduction risk dying before the next fire. Alternatively, plants that allocate resources to longer life span may be burnt before they have become reproductive. What is the optimal delay between germination and first flowering (maturation time) in plants with these life histories?

Clark (1991) has recently addressed this question for typical fire life-histories. Unlike most life-history theory, which assumes continuous recruitment (Caswell, 1982), Clark explicitly recognized the two-phase dynamics of episodic recruitment and subsequent cohort attrition typical of fire life-histories (Chapter 4). He considered two general life-history types:

- Type A – equivalent to fire-recruiting non-sprouters, where the disturbance that initiates regeneration also results in death of the mature plant. He assumed that the optimal life history is that which maximizes the *probability that a plant will be reproductively mature at the time of the next fire*. This differs from analyses of continuously growing populations which explore ways of maximizing net reproductive rate (e.g. Stearns, 1977; Caswell, 1982).
- Type B – equivalent to fire-recruiting sprouters, where mature plants survive the disturbance that produces regeneration opportunities. Type B plants include those with a vulnerable juvenile period when plants can be killed by a disturbance. Clark assumed that the optimal life history here is that which maximizes the *number of recruitment episodes experienced over the reproductive life of the plant.*

These optimality criteria are more explicit than conventional theory and closer to the biological reality of episodic recruitment in fire-recruiting species (Chapter 4). They assume that seed production that does not coincide with recruitment episodes is essentially wasted (although accumulation of seeds in seedbanks means that wastage is not complete).

Optimal maturation time for a given fire regime depends on the relationship between mortality patterns between fires, the duration of the reproductive period and the probability of fire as a function of stand age.

Mortality patterns influence the chances of surviving from one fire to the next. Clark assumed constant cohort mortality after initial establishment thus:

$$S(t) = e^{-\lambda t}$$

where $S(t)$ is survival at time t and λ is per capita mortality rate.

The duration of the reproductive period (from maturity to death) is assumed to be determined by the trade-off between reproduction and life span (Loehle, 1988). Loehle's analysis suggests a direct proportionality between maturation time, t_1, and the duration of the reproductive period R_α calculated as:

$$R_\alpha = t_1(\alpha - 1)$$

where $\alpha = 5.4$ for angiosperms and 6.3 for gymnosperms. These values are derived empirically from the reciprocal of the regression slope relating maturation age to longevity in Loehle's (1988) analysis of North American trees. They may differ for other data sets.

Finally, the probability of fire in relation to the time elapsed since the last one is needed to predict the likely success of any combination. For example, if fire return intervals are long, plants would need longer life spans to survive before a fire triggers the next recruitment episode. Clark used a variety of different probability density functions, including Weibull distributions derived from fire scars on *Pinus resinosa*, to predict the likelihood of fire with increasing time since the last burn (Chapter 2).

From these three relationships, equations can be developed that predict optimal maturation time for different patterns of disturbance and inter-fire mortality for sprouting and non-sprouting life histories. For example, if the probability of fire increases exponentially with increasing time since the last fire, but the probability of burning is constant across all ages of vegetation, the optimal maturation time for a non-sprouter is given by:

$$t_1^* = \frac{A}{\lambda_s + \lambda_p}$$

where t^* is optimal maturation time, A is a constant relating to reproductive period, λ_s is the probability of disturbance and λ_p is the (constant) thinning rate. This gives a quantitative estimate for the qualitative prediction that a short maturation time would be optimal where there is a high probability of disturbance and/or where there are high rates of inter-fire mortality.

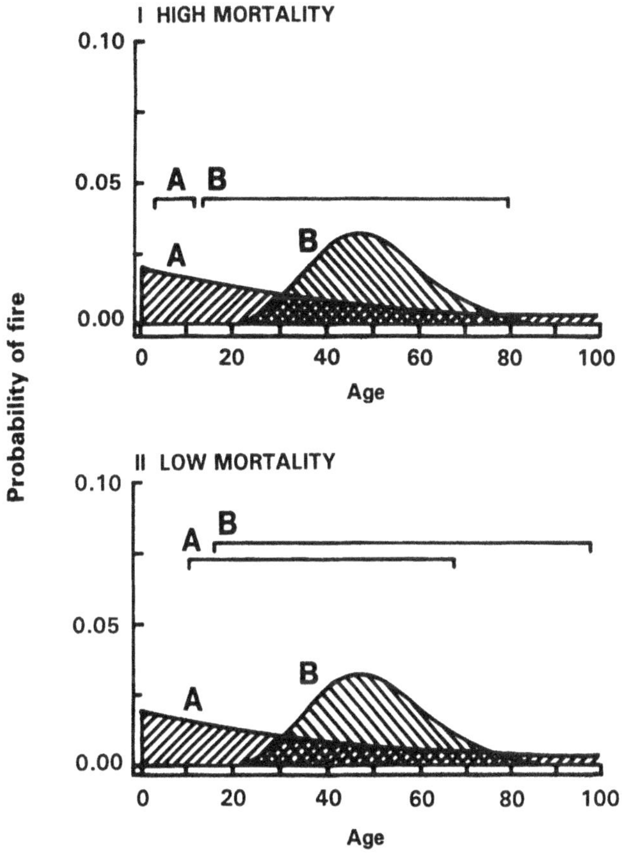

Figure 6.6 Effects of disturbance on optimal life histories for non-sprouters (from Clark, 1991). The optimal reproductive period, indicated by the horizontal lines, depends first on the probability of disturbance: A, constant (= exponential distribution) and B, increasing through time (e.g. due to fuel build-up); and secondly on mortality rate between fires: I, high mortality and II, low mortality. Note that optimal reproductive period is very sensitive to the type of disturbance regime at high mortality rates (I, optimal maturation age of 4 years for A, 17 years for B) but there is extensive overlap at low mortality rates (II).

Optimal maturation time changes with changing patterns of disturbance, for example, if fuel accumulates with age. It also changes if inter-fire mortality is high (Figure 6.6).

Similar calculations for sprouting species predict that optimal maturation time should be greater than for non-sprouters under the same disturbance regime and if they share the same interfire mortality.

Clark tested predictions of his model using data from populations of *Pinus resinosa*, a sprouter, and *P. banksiana*, a non-sprouter (Figure 6.7). The two species co-occur and therefore share identical fire regimes yet have very different maturation times. As predicted, *P. resinosa* begins

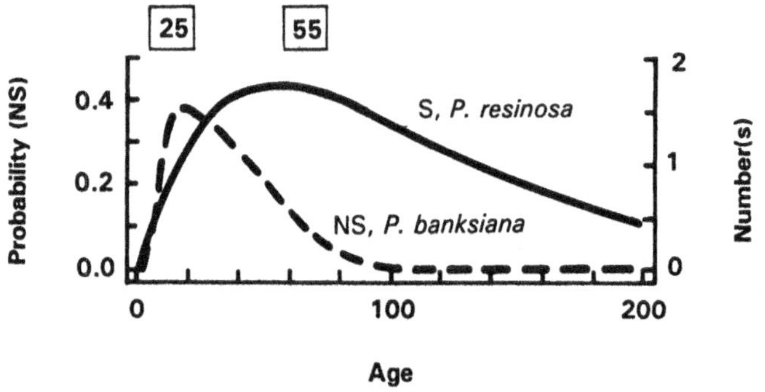

Figure 6.7 Comparison of optimal life histories where probability of reproduction is maximised in a non-sprouter (NS, *Pinus banksiana*) and number of reproductive episodes in a co-occurring sprouter (S, *Pinus resinosa*). Both species experience the same disturbance regime with an average interval of 60 years and an increasing probability of disturbance since the last fire. The predicted optimal age of maturation is close to the observed shown in the boxes (from Clark, 1991).

reproducing much later than *P. banksiana*. The predicted and observed maturation times coincide closely providing some support for the argument that the two life histories are maximizing different quantities: number of reproductive episodes in sprouters, and probability of being reproductively mature at the time of the next fire in non-sprouters.

Clark's model is appealing in combining disturbance and plant population dynamics to explore optimal life histories. The explicit recognition of two-phase dynamics of plants is far more realistic than the assumption of continuous recruitment in conventional life-history analyses. The argument that different quantities are optimized seems reasonable but needs much wider testing against empirical data from other species and systems. It would also be useful to extend the model to species with seedbanks.

Finally, to return to the original question of why plants senesce and die if intervals between fires are too long: maturation time and longevity may have evolved under a fire frequency of shorter mean fire intervals than the one currently experienced. In general, senescent populations occur in areas managed for fire suppression. It would be interesting to test this life-history optimization model, and the trade-off arguments, in the reverse situation where fire frequency has increased. Arson is a favourite urban pastime. The model predicts that populations in high-frequency burn areas should flower earlier than those burnt under 'natural' conditions. On islands, where fires are rare, flowering should be delayed relative to mainland populations.

6.4 TIMING OF REPRODUCTION AND RECRUITMENT

As noted in Chapter 4, the two dichotomies of fire life-histories are between sprouters and non-sprouters on the one hand, and fire-recruiting and non-fire-recruiting regeneration on the other. We have dealt with sprouter/non-sprouter evolution above. Now we consider the evolution of fire-stimulated reproduction and regeneration. These include all aspects of reproduction that lead to the timing of seedling emergence after a burn, such as fire-stimulated flowering, fire-induced germination of soil-stored seedbanks, and fire-induced seed release from canopy-stored seedbanks (Chapter 3 gives examples). Fire-recruiting species often co-occur with non-fire-recruiting species (Keeley, 1986; Manders, 1990). For example, woody plants with short-lived fleshy fruits and seeds that recruit between fires occur in Californian chaparral, Mediterranean maquis, South African fynbos and other fire-prone shrublands and woodlands. They typically replace species with soil- or canopy-stored seedbanks, and fire-stimulated recruitment, if there are long fire-free intervals (Keeley, 1986; Manders, 1990).

What circumstances favour the evolution of fire-stimulated reproduction? Plants which recruit only after fire must produce as many offspring in this very brief period as the sum of all progeny produced between fires by non-stimulated species. Thus a fire-stimulated grass that flowers only after fire may produce 10 times more seeds than a season's production of a non-stimulated variant. However the difference would be trivial when discounted over decade-long fire intervals. There is an added cost to fire-stimulated flowering since the post-burn progeny must equal the progeny of those establishing between fires. These costs are not unique to fire-recruiting species. They are characteristic of all plants with episodic reproduction, including mass flowering (Augspurger, 1984) and mast fruiting trees and shrubs (Waller, 1979; Silvertown, 1980).

6.4.1 The evolution of serotiny

Studies of delayed reproduction have focused mainly on the particular benefits, such as predator and competitor avoidance, which are derived from reproducing/recruiting in post-burn conditions. A few studies have looked at the broader questions of the effect of disturbance regime on the evolution of delayed reproduction in general and seedbanks in particular (e.g. Givnish, 1981; Muir and Lotan, 1985). The most successful of these concern the evolution of serotiny. Serotiny (Chapter 3) is the accumulation of a seedbank in the canopy of a plant through delayed seed release. Because the presence and longevity of serotinous seedbanks are much easier to measure than soil seedbanks, they provide almost unique examples of the evolution of seedbanks in relation to disturbance regimes (for a recent review see Lamont *et al.*, 1991b).

Serotiny varies within populations, between populations and between related species. This variation is closely linked to fire patterns. Lodgepole pine, *Pinus contorta*, is one of the most wide-ranging and variable of western North American conifers. *P. contorta* var. *latifolia* has serotinous forms (with cones that require heat to melt the resin which bonds cone scales

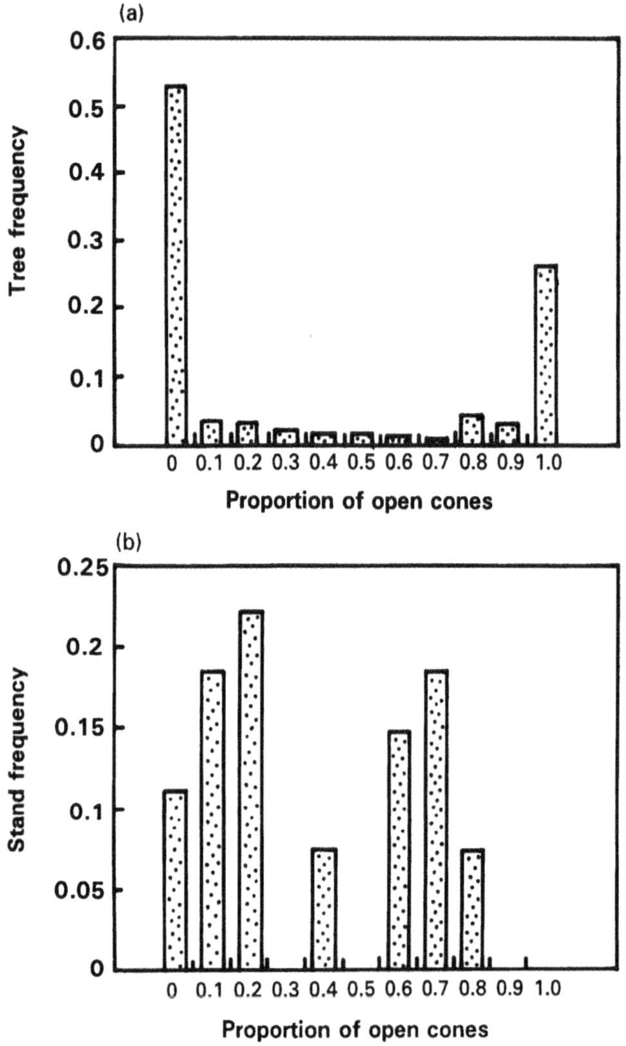

Figure 6.8 Frequency distribution of open (as opposed to serotinous) cones in lodgepole pine, *Pinus contorta*. (a) Trees more than 50 years old; (b) number of stands. (Data from Muir and Lotan, 1985.)

together) and open-coned forms with cones that open at maturity. Most trees are either closed-coned or open-coned and most populations are dominated by trees of one cone type or another (Figure 6.8) (Muir and Lotan, 1985). Cone type, and therefore the presence of a seedbank, is genetically controlled, possibly by a single locus (Teich, 1970; Perry and Lotan, 1979). Muir and Lotan (1985) studied the frequency of cone type in relation to the type of disturbance initiating a stand and the estimated fire frequency. Closed cones predominated wherever stands were initiated by large, stand-replacing burns, and open cones where trees had established in tree-fall gaps and not after fire (Table 6.4). Mean fire frequency was a poor predictor of the dominant cone type. The cone polymorphism appears to be

Table 6.4 Relationship between frequency of serotinous (closed) cones in *Pinus contorta* and type of disturbance in Montana (Muir and Lotan, 1985). Differences are significant (ANOVA, $P < 0.001$)

	Type of disturbance		
	Non-fire	*Partial burn*	*Stand-replacing burn*
Proportion serotinous	0.26	0.51	0.73
Standard error	(0.08)	(0.10)	(0.05)

retained, in their Montana study area, by a disturbance regime which is highly variable in time and space, with substantial gene flow between adjacent populations (Muir and Lotan, 1985).

This pattern of variation suggests that serotiny evolves when:

- stand-killing fires burn over extensive areas;
- interfire intervals are too short for much reproduction to occur from plants establishing between fires;
- fire size is too large for significant seed dispersal from adjacent unburnt areas (McMaster and Zedler, 1981).

The rationale is that trees with serotinous cones will have more seeds than open-coned trees for recruitment after a fire because of the accumulated seedbank. However, because seed release is delayed until fire, they will disperse fewer seeds than open-coned trees into gaps created by single tree-falls. The second consideration relates to the cost of deferred reproduction discussed above. Finally, fire size should be large relative to dispersal distance, or serotinous seeds would have to compete with seeds from open-coned trees invading from the unburnt periphery (McMaster and Zedler, 1981).

There is a general association of serotiny with large crown fires. Open-coned or weakly serotinous variants are associated with sites with low fire frequencies, less-intense fires which do not reach the canopy, and establishment of individuals between fires. These patterns have been reported for northern-hemisphere conifers (Givnish, 1981; McMaster and Zedler, 1981; Borchert, 1985; Muir and Lotan, 1985) and southern-hemisphere conifers and angiosperms (Cowling and Lamont, 1987; Lamont *et al.*, 1991b). Given the comparative ease with which serotiny can be assessed compared to soil-stored seedbanks (Plate 2), they are the obvious choice for empirical studies testing general models of seedbank evolution. So far the availability of conceptual (and especially analytical quantitative) models, rather than data collection, has limited progress.

Soil-stored seedbanks are common in fire-prone environments but, as in common with other environments, the technical problems of quantifying seedbanks have largely excluded evolutionary studies (Parker and Kelly, 1989).

6.4.2 The selective advantages of fire-stimulated reproduction

The common occurrence of fire-stimulated flowering (Plate 1), seed germination and seed release implies substantial growth advantages in post-burn conditions. Many studies have compared burnt and unburnt populations in terms of growth rates, flower production, seed set, seed predation and seedling survival. For example, seed predation and seedling growth and survival was studied in *Eucalyptus delegatensis*, alpine ash, after fire (O'Dowd and Gill, 1984). Fire stimulated a massive, synchronized seed release from the serotinous capsules of the eucalypt. The seed rain satiated the ant seed predators, causing a reduction from 65.3% pre-fire seed removal to 13.6% post-fire removal for a few weeks after the burn. Rather surprisingly, conditions after the fire did not improve early seedling growth, levels of herbivory or post-burn survival. O'Dowd and Gill's study is typical of many showing that seed predation is reduced after fire – though probably more because predator densities are reduced by burning than from satiation by massed seed release (Bond, 1984; Le Maitre, 1984; van Hensbergen *et al.*, 1992). Predator satiation due to fire-stimulated mass flowering and fruiting has also been invoked to explain mass flowering in tropical plants (Augspurger, 1984) and mast fruiting in temperate and tropical trees (Silvertown, 1980; Sork, 1983).

Predator satiation is only one of several possible explanations for synchronous reproduction (Kelly, 1994). In some studies, fire-induced flowering had little effect on seed predation. In the Australian cycad, *Macrozamia communis*, cone production was stimulated by fire, but subsequent seed dispersal by possums was poorer in the masting populations and seed predation higher (Ballardie and Whelan, 1986). Here fire may act rather as a cue to synchronize reproduction in the population and so increase the chances of successful pollination (Ballardie and Whelan, 1986). Similar arguments have been proposed for the cueing of mass flowering in forest trees by unusually dry weather conditions (Norton and Kelly, 1988; Kelly, 1994).

While some form of reproductive advantage has often been shown in comparisons of burnt and unburnt fire-stimulated populations, there have been few attempts to assess these critically – quantitatively – over a fire cycle. For example, O'Dowd and Gill (1984) reported a fivefold advantage for seeds dispersing after fire compared to *a single dispersal event* between fires. Clearly, if fire intervals were much longer than 5–10 years, the fitness advantage of serotiny would fall away. We feel there is still much to be learnt about selection favouring deferred reproduction from a study of fire-recruiting life-histories.

6.4.3 Comparisons among fire-recruiting reproductive modes

What explains the variability in modes of fire-stimulated reproduction? Why, for example, do some species develop canopy seedbanks and others soil-stored seedbanks when both are cued by fire? Clearly 'history' in the form of pre-existing woody cones, may have pre-disposed conifers to evolve serotiny rather than hard-seededness to form seedbanks. However, there are instances of soil-stored seeds evolving from serotinous ancestors in the Cape Proteaceae (Midgley, 1987) and distribution patterns of confamilial taxa with the different seedbank types seem to differ (Le Maitre and Midgley, 1992). There is growing interest in the possible selective pressures favouring different modes of fire-stimulated reproduction. For recent reviews on ideas suggested to date, the reader is referred to Lamont *et al.* (1991b) and Le Maitre and Midgley (1992).

6.5 EVOLUTION OF FIRE-SURVIVAL TRAITS

Many plants survive fires vegetatively through protection from thick bark, persistent leaf bases or from underground buds insulated by soil, but is there any evidence that these traits have evolved in response to fire? It is difficult to attribute survival traits uniquely to fire since most have multiple benefits, including protection against herbivory, frost and pathogens. The evolution of grasses, and their capacity to sprout from leaf bases insulated by old leaf sheaths, heralded one of the most profound changes of life on Earth (Stebbins, 1981; Axelrod, 1985). Stebbins (1981) has emphasized the coevolution of grasses and large grazing herbivores. However, fire, rather than herbivores, would have acted on grass fuels to kill trees and open up vast tracts of grasslands and savanna, creating an entirely new selection regime in which herds of grazing mammals could thrive. But the leaf sheaths that insulate against fire also provide protection from frost and drought, and the relative selective effects of fire, drought and frost on grass morphology have not been disentangled.

Savanna trees usually have thick bark but whether this is a fire adaptation is open to question. *Pinus merkusii*, the only pine with a natural distribution range that crosses the equator, shows intraspecific variation in bark that may be due to the selective effects of fire (Stott, 1988; Stott *et al.*, 1990). The savanna form of the pine, on the mainland of South-East Asia, has a thick bark, shorter stature and a seedling 'grass' stage, similar to the long-leaf pine of the eastern United States. On the island of Sumatra, where fires are rare, the species has much thinner bark, is much taller and lacks a grass stage. During the grass stage (lasting up to 7 years), the plant has a strong rootstock and is remarkably resistant to ground fires. There is a clear trade-off between survival and growth in the seedling stage, exploited by foresters who prefer populations with no grass stage because of their faster growth rates. Comparisons of fire survival 'adaptations' between savanna trees and sister species in forests that do not burn would offer interesting insights into the selective effects of fire.

6.6 EVOLUTION OF FLAMMABILITY

6.6.1 The Mutch hypothesis

Many vegetation types depend on fire for their continued persistence. Without fire, these communities are replaced by non-flammable forests and scrub (Chapter 7). Mutch (1970) was among the first to argue that fire-prone vegetation may have evolved characteristics which increase flammability and thereby contribute to the perpetuation of fire-dependent communities. He argued that though ignition may occur 'accidentally or randomly, the character of burning is not random. . . . Fire dependent plant communities burn more readily than non-fire-dependent communities because natural selection has favoured development of characteristics that make them more flammable.'

Mutch's hypothesis has been widely cited by fire ecologists since it seems to provide an evolutionary rationale for the differences between fire-prone and fire-free communities. For example, studies on pines and oaks have shown that pine litter is more flammable than oak litter and that pines require fire to avoid successional replacement by oaks. This has been taken as evidence that flammability has evolved in pines as part of its competitive armoury (Williamson and Black, 1981; Rebertus *et al.*, 1989). Greater flammability in fire-tolerant vegetation has also been reported for eucalypt woodlands (Mount, 1964; Jackson, 1968; Dickinson and Kirkpatrick, 1985), arid shrublands (Buckley, 1984) and chaparral (Philpot, 1969). Contrary evidence, showing little or no correlation between fire regime and flammability, has been reported by Bowman and Wilson (1988) in Australian tropical vegetation, while van Wilgen *et al.* (1990a) reported that differences in flammability between fire-prone fynbos and fire-free Afro-temperate forest were due to differences in moisture content and fuel packing, rather than chemical differences influencing flammability.

6.6.2 The problem with the Mutch hypothesis

Despite its wide appeal, the Mutch hypothesis is flawed (Snyder, 1984; Troumbis and Trabaud, 1989). Although communities that burn more readily may be more flammable than those that do not, the differences in flammability may be purely coincidental. Mutch presented his hypothesis at the level of species or community. Thus pines are assumed to have evolved flammability to avoid replacement by oaks. However, natural selection operates on individuals and it is hard to imagine the individual benefits of burning. Flammability itself is a difficult concept to define in relation to evolution by natural selection because it is influenced by many different and apparently unrelated plant properties (Troumbis and Trabaud, 1989). At the level of the individual plant, chemical traits that enhance flammability include high cellulose (rather than lignin) content of leaves, and high levels of ether-extractive secondary compounds, such as terpenes and waxes (Rundel, 1981). Structural traits that enhance flammability include high surface area to volume ratios (fine foliage, thin branches), branching patterns that carry fire from the surface to the canopy, and scleromorphic leaves (low moisture content) (Rundel, 1981; Papio and Trabaud, 1991). Both sets of traits are likely to have other effects on plant function, such as deterring herbivores, nutrient retention, drought tolerance or faster growth rates (Snyder, 1984; Troumbis and Trabaud, 1989). Untangling these alternative functions from flammability is not an easy task. Quite similar difficulties confront attempts at explaining plant secondary chemicals as antiherbivore adaptations (e.g. Edwards, 1989).

6.6.3 Individual benefits of burning

Can the Mutch hypothesis be re-phrased to argue for individualistic selection for enhanced flammability? Imagine a tree which survives low-intensity burns but is killed by crown fires. Under what conditions would a flammable mutant (a 'torch' – perhaps one that produces a lower, more densely branched canopy) spread in a less-flammable population ('damps')? After a fire there will be patches of burnt 'torches' and unburnt 'damps' (Figure 6.9). Seedlings of 'torches' would be able to invade burnt, but not unburnt, patches. However, they would have to compete with seedlings of 'damps' which also recruit into the burnt patches.

Bond and Midgley (1995) recently explored this kind of scenario using models that simulate fire mortality, dispersal into gaps created by such mortality, and subsequent seedling competition. In their model, flammability could not evolve if only 'torches' burnt because the gaps would be swamped by seeds from unburnt 'damps'. However, in dense populations, crown fires are likely to spread from 'torches' to neighbouring 'damps'. This neighbourhood effect of flammability proved necessary, but not a sufficient condition, for 'torches' to spread. Additional fitness benefits, such as more competitive seedlings or greater fecundity, were needed to displace the 'damp' genotype. For plants with some chance of resprouting after crown fires, the neighbourhood effect alone was sufficient for flammability to evolve. However, 'torches' were more likely to spread with additional fitness benefits.

One prediction arising from this analysis is that traits that enhance flammability in trees or tall shrubs are most likely to occur in dense populations. Solitary conifers or sparse savanna trees, for example, should be low on the flammability scale – gregarious pines or Mediterranean shrubs should be high.

Figure 6.9 Conditions under which enhanced flammability (shaded) might be an advantageous trait and spread through an ancestral less-flammable (unshaded) population. Enhanced flammability can create recruitment opportunities if it causes a plant to burn and the fire to spread to less-flammable neighbours. Flammability is more likely to spread if flammability traits carry additional benefits, such as faster growth rates.

Bond and Midgley (1995) suggested that many of the structural and chemical traits that enhance flammability provide the needed additional benefits for flammability to evolve. For example, low, densely-branched crowns enhance growth in early successional environments but also increase the risk of crown fires. Unfortunately, this synergic effect makes it even more difficult to separate the role of fire. One approach is to explore variation in properties that influence flammability but seem to have no other function. For example trees and shrubs vary in the tendency to shed dead branches. Bond and Midgley predicted that flammable variants should retain dead branches and occur in dense stands whereas less flammable variants should be sparse and undergo self-pruning.

This neighbourhood model does provide one set of conditions under which flammability could evolve for individual benefit. However, the evolution of flammability has ramifications far beyond individual advantage. If just a single common species evolved properties that increased the probability of burning, it could change fuel properties of the vegetation, leading to the disruption of entire communities. Fire appears to act as an ecological filter, permitting access only to those species compatible with the disturbance regime (Chapter 7). In this sense, it may be a key agent in the evolution of whole floras (Chapter 9).

6.7 CONCLUSIONS

With fire being so ubiquitous on the planet, it is surprising how little rigorous attention has been given to its evolutionary consequences. Fire life-histories provide an interesting counterpoint to many traditional concerns of evolutionary ecology. For example the cost of reproduction seems less significant than the cost of survival. In the majority of cases, fire survival by sprouting or production of thick bark is the ancestral condition. The cost of allocation to survival tissues leads to reduced juvenile growth and reproduction. Reduced allocation to survival brings substantial fitness advantages for non-sprouters. Physiological studies of costs and allocation trade-offs are in their infancy. We could find no reference, for example, to the costs of bark production. Are there trade-offs in seedling growth and survival linked to bark thickness in fire-prone conifers, eucalypts and savanna trees?

Formal models exploring fire life-histories are also in their infancy, though those of Clark (1991) are a major contribution. At least we are moving away from the simplistic vertebrate-based models of the past and can justifiably ignore unusable concepts such as r and K selection (is a fire ephemeral herb any more 'r' selected than a non-sprouting tree when both share the same generation time?).

We have not attempted to define 'adaptation'. However, the word has been ill-used in many discussions of plant traits in fire-prone environments. Cladistic methods offer much promise for tracking the historical sequence in which various traits evolved. For example, Midgley (1987) has shown the power of the phylogenetic perspective in determining which traits are derived and which ancestral for a group of fire-prone woody shrubs (see also Linder, 1995). What does seem to be clear is that repeated combinations of traits, adaptive syndromes, do occur in widely separated geographic regions and distantly related taxa. There are patterns – of sprouters, non-sprouters, reproductive synchrony and flammability – and a good start has been made in explaining them.

7 Fire, competition and the organization of communities

'Ever since Darwin, competition has been considered to be one of the major forces shaping the morphology and life history of plants and the structure and dynamics of plant communities' (Grace and Tilman, 1990).

7.1 INTRODUCTION

Thus far, we have been considering only species in isolation. How does fire influence mixtures of species? Community ecology, the study of species assemblages, has a large and increasingly complex set of explanations for patterns of species distribution and abundance. We contrast these explanations, many based on the importance of competition, with observations on communities by fire ecologists. Keddy (1992) set out the broad aims of community ecology as searching for assembly rules, which predict which kinds of species fit together to make a community, and response rules, which tell us how communities might change given some particular perturbation or changing environment. There is growing interest, too, in how the rules change depending on the scale of interest. Fire ecologists seem to have expended their efforts in roughly inverse proportion to mainstream plant community studies: we know a good deal about response rules but rather little about rules of assembly. However, fire ecologists were among thc first to appreciate the importance of scale and have initiated pioneering studies in landscape ecology.

To limit the scope of this chapter, on a large and challenging topic, we summarize the key questions of community ecology as follows:

- What determines changes in plant communities over time?
- What determines changes in plant communities over spatial gradients?
- Which combinations of species occur together and why?
- What are the determinants of species diversity?
- How do processes influencing communities vary in importance in different parts of the world?

Answers to these questions are fundamental for providing assembly and response rules which help us understand, and manage, natural communities. In this chapter, we discuss briefly the main ideas of current theory, in which competition usually plays a central role. In fire-prone communities, much greater emphasis has been placed on the interplay between the disturbance regime and plant life-history attributes in seeking response rules. This is essentially a non-interactive view of community dynamics. Vital attributes relating to fire survival, reproductive timing, longevity and recruitment are seen as the key to understanding vegetation dynamics. These are seen as absolute species properties which exist regardless of the competitive milieu. Dynamics driven by competitive interactions, in which relative growth rates and resource capture are of central concern, have been largely ignored. We discuss the strengths, and limitations, of the vital attribute approach in developing community rules for different fire-prone communities. We discuss briefly the rapidly expanding field of landscape ecology, where fire studies have made pioneering contributions. At landscape scales, fire has worked synergistically with climate, soils and humans to shape whole biotas. This interactive web has suffered extreme disruption in the twentieth century. We return to the consequences of this disruption in Chapter 9.

7.2 WHAT DETERMINES CHANGES IN PLANT COMMUNITIES OVER TIME?

It has long been recognized that plant communities change over time. Attempts to understand the patterns and underlying processes of succession have been a central focus of plant ecology (recently reviewed by Pickett *et al.*, 1987; Luken, 1990; Glenn-Lewin *et al.*, 1992). The dominant view until the 1970s was developed by Clements (1916) who saw succession as a relatively orderly, predictable approach to a dynamic equilibrium. After some initial disturbance, plants colonize the space released. Pioneer species facilitate the invasion of later species, for example by modifying the soil, or are ousted by a sequence of species of successively greater competitive ability. Ultimately an equilibrium is reached where, if undisturbed, only top competitors persist. This equilibrium, the climax, was assumed to be self-sustaining and persistent.

Ecologists now recognize much greater diversity in successional patterns and processes (Connell and Slatyer, 1977; Pickett *et al.*, 1987). However, the idea of a predictable sequence of species replacement, the **relay floristic** model of succession, is still central to much ecological thinking. It has been developed into a logical protocol for predicting community dynamics based on competition as the central process. The idea is to rank species according to competitive ability from observation or experimental analysis of interactions between species pairs. This provides the trajectory of expected species replacement sequences. Given sufficient time, succession should proceed until the most competitive species dominate the community. These equilibrial predictions are then compared with the field situation.

Communities may often be held at a state below the anticipated competitive equilibrium so that subsidiary processes which block the relay from reaching its competitive end point may need investigating. The final step is to modify the successional model to accommodate these processes. In this system, resource competition is the engine driving succession, while other processes, such as herbivory, merely accelerate or brake it. Even if equilibrium is never observed, the predicted replacement sequence is a powerful framework for identifying the pathways and direction of temporal change.

A key requirement in this protocol is a competitive ranking of species measured by experimental or descriptive studies (Keddy, 1989). Elegant mechanistic models of competition for resources have been developed successfully to identify traits which order species along temporal replacement sequences (Horn, 1971; Bazzaz, 1979; Grime, 1979; Tilman, 1982, 1988). Tilman has run a comprehensive long-term experiment which exemplifies the competitively driven successional model in old field communities in Minnesota (Tilman, 1990; Wedin and Tilman, 1993). He used greenhouse experiments to confirm the ranking of relative competitive ability for a number of grass species, based on the nutrient requirements of each species in monoculture. Observed patterns of old field succession partly supported the successional model, but the top competitor did not reach its predicted dominance in the field. Tilman attributed the discrepancy to a subsidiary process – colonization – which, in his system, was inversely related to competitive ability (Tilman, 1994). Top competitors were slow colonizers and their entry to the system was inhibited by the residue of litter produced by earlier site occupants.

7.2.1 Fire succession

The Clementsian concepts of relay floristics and a self-sustaining climax were enthusiastically applied to post-fire succession (e.g. Phillips, 1934–35) but quickly proved inadequate. A very different set of community response rules was developed by fire ecologists to explain successional patterns, especially in woody vegetation. These emphasize the interplay between disturbance and the timing of key life-history events. Competition is virtually ignored. There are a great many descriptive accounts of secondary succession after fire. We summarize a few of them to show how the patterns have helped to formulate thinking, and then discuss alternative models that have been developed to predict vegetation dynamics after fire.

7.2.2 Mediterranean shrublands

Succession after fire in the chaparral of southern California was described in a landmark paper by Hanes (1971), which represented a significant challenge to the dogma of the time. Much of chaparral is dominated by a single species, chamise (*Adenostoma fasciculatum*). Hanes described post-fire succession in chamise chaparral as 'unusual'. He contended that 'chaparral succession was set apart from the usual sequence of change that characterizes secondary succession in most plant communities', because the shrubs composing the mature community were present in the vegetation in the first year after fire. Subsequent colonization of the site by new species was very limited. Hanes suggested the term 'autosuccession' to distinguish self-replacing vegetation like chaparral from types that must be preceded by distinct seral stages. Chamise chaparral stands older than 60 years after fire were described as 'senile, ... trashy and unproductive' (Hanes, 1971); the vegetation was not a true climax vegetation because it could not perpetuate itself in the absence of fire, and there was little evidence to support claims that chamise-chaparral would eventually be succeeded by a live-oak woodland. This, together with evidence for the production of phytotoxic substances in old stands of chaparral, was used to spearhead an attack on the policies of fire exclusion prevalent at the time of Hanes' paper.

Very similar successional patterns have been reported for French garrigue, a shrubland dominated by a resprouting scrub oak, *Quercus coccifera* (Trabaud and Lepart, 1980). Here floristic richness reached a maximum during the first 2 years after fire, decreasing thereafter and becoming stable after 5–7 years in terms of species present. Early species include annuals and other ephemerals that come from outside the communities, and disappear as the stands age. Trabaud and Lepart spoke of the 'terminal community' (an equivalent of Clements's climax). Most species of the terminal community appeared quickly after fire, which led them to the observation that 'there is no real secondary succession, in that species or communities do not follow each other after fire', but rather a rapid reappearance of species belonging to the pre-burn community. The floristic richness of communities at 10–15 years after fire (the limits of the study) did not differ markedly from more mature communities in the area. In both examples, initial floristic composition is highest early in succession and subsequent changes are due merely to the progressive loss of species with short life spans or suppression of species by the overstorey.

7.2.3 Scottish heathlands

Heathlands in Britain (and elsewhere in Europe) are dominated by the ericaceous shrub *Calluna vulgaris*. A knowledge of the life cycle of *Calluna* forms an important basis for the management of heathlands, and was developed from the observations of Watt (1947, 1955, 1964). Following establishment after fire, *Calluna* passes through four distinct phases during its life history (pioneer, building, mature and degenerate, followed by death). Fire is used in heathland management to interrupt the ageing process of *Calluna*, keeping it in a young, productive and nutritious state and increasing potential herbivore production, and is also used to prevent the transition from heathland to scrub or woodland (Hobbs and Gimingham, 1987). For example, while fire will enhance the establishment of tree species such as

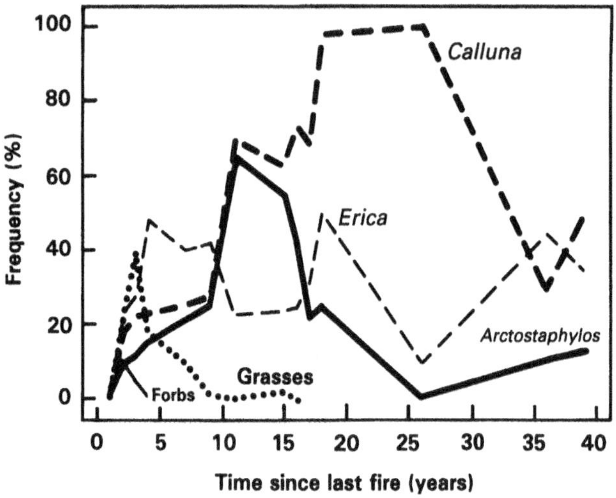

Figure 7.1 Post-burn succession of vascular plants in *Calluna vulgaris* heathland in Scotland inferred from percentage frequency in a chronosequence (from Hobbs *et al.*, 1984b). Note that succession occurs by selective elimination from the initial group of taxa present after the fire.

Pinus sylvestris and *Betula* in heathlands, rapid regrowth of *Calluna* on many sites ensures that these seedlings do not survive. Many herbaceous species die out under closed canopies of *Calluna*, and seed stores are rapidly depleted (Figure 7.1). Those seeds that do germinate fail to establish under the closed canopy. However, succession is complicated by several factors, such as the composition of the pre-fire community, fire severity, weather patterns in the post-fire period and the activities of herbivores (Hobbs and Gimingham, 1984). These factors can produce multiple pathways of succession and very different post-burn communities. Fire intervals of 12–15 years provide the optimum regime for maintaining community diversity.

7.2.4 Eucalyptus wet-sclerophyll forests

The giant wet eucalypt forests of Australia represent an extreme physiognomic contrast to these shrublands. These forests are dominated by some of the tallest trees in the world (50–110 m). Incredibly, these eucalypts are fire-recruiting non-sprouters (Chapter 4) and cannot regenerate unless burnt in crown fires of awesome intensity at intervals of 200–300 years (Jackson, 1968; Ashton, 1981). Despite the physiognomic differences, post-fire succession in these forests has many features comparable to fire-prone shrublands (Ashton, 1981). Like British heathlands, succession can follow multiple pathways depending on fire history. Immediately after fire a luxuriant growth of herbs is common, together with the seedlings of eucalypts and other woody species which survived burning in canopy or soil-stored seedbanks and which will eventually form the overstorey. Species richness is greatest in the first decade after fire and declines as the stand ages. After the first initial flush of fire weeds, the course of succession depends for many decades on the growth of the post-burn assemblage – the **initial floristic composition** (Egler, 1954). If the fire return interval is very long, and if seed sources are near and soils are fertile, rainforest species begin to invade. Community dynamics are controlled by fire. Short fire intervals can eliminate the dominant eucalypts by killing them before they have accumulated adequate seedbanks, so transforming the community into a highly flammable heath shrubland. Intermediate fire frequencies maintain less flammable, but still fire-dependent, eucalypt woodlands. Very long intervals result in the replacement of these wet sclerophyll forests by non-flammable temperate forests whose shaded understorey inhibits recruitment of the light-demanding canopy eucalypts. The alternative pathways of succession, and alternative states of wet sclerophyll vegetation, are primarily driven by fire return time (Jackson, 1968; Ashton, 1981; Figure 7.2). Successional patterns are less predictable than in the case of garrigue or chamise-chaparral and highly dependent on the timing and intensity of fire. Similar examples of alternative flammable and non-flammable vegetation 'states' occurring under the same environmental conditions are common in many parts of the world (see below).

These examples of post-fire succession differ greatly from the old field succession pattern typifying Clementsian succession (Figure 7.3). The first obvious divergence is that there is no progression to a self-sustaining climax. Succession is cyclical under the same fire regime or results in multiple pathways to alternative communities if the fire regime changes. Very long return intervals can result in replacement of flammable vegetation by an alternative stable state of non-flammable vegetation. Secondly, there is little evidence for competitively driven replacement sequences. Species richness is highest soon after the fire and succession proceeds by successive elimination of species. The order of elimination is determined by relative longevity. The dominance of each species is transitory as it ages and dies. Losses are not permanent since seeds usually persist in seedbanks and the plants reappear after the next fire. The 'engine' driving successional change is the different life histories of the species. The implication is that, unlike Tilman's grasses, additions or deletions of species will not change the successional outcome for the remaining species. The key to predicting successional change is the timing of life-history events in relation to disturbance, not position in the competitive hierarchy.

7.2.5 Predicting the paths of fire succession

If relative competitive ability is a poor predictor of temporal change in fire communities, what are the alternatives? Noble and Slatyer (1980) proposed a minimum set of vital attributes necessary to predict successional change in communities (Box 7.1). Their system makes explicit a general view of fire ecologists – that temporal change is driven by relative longevities of species

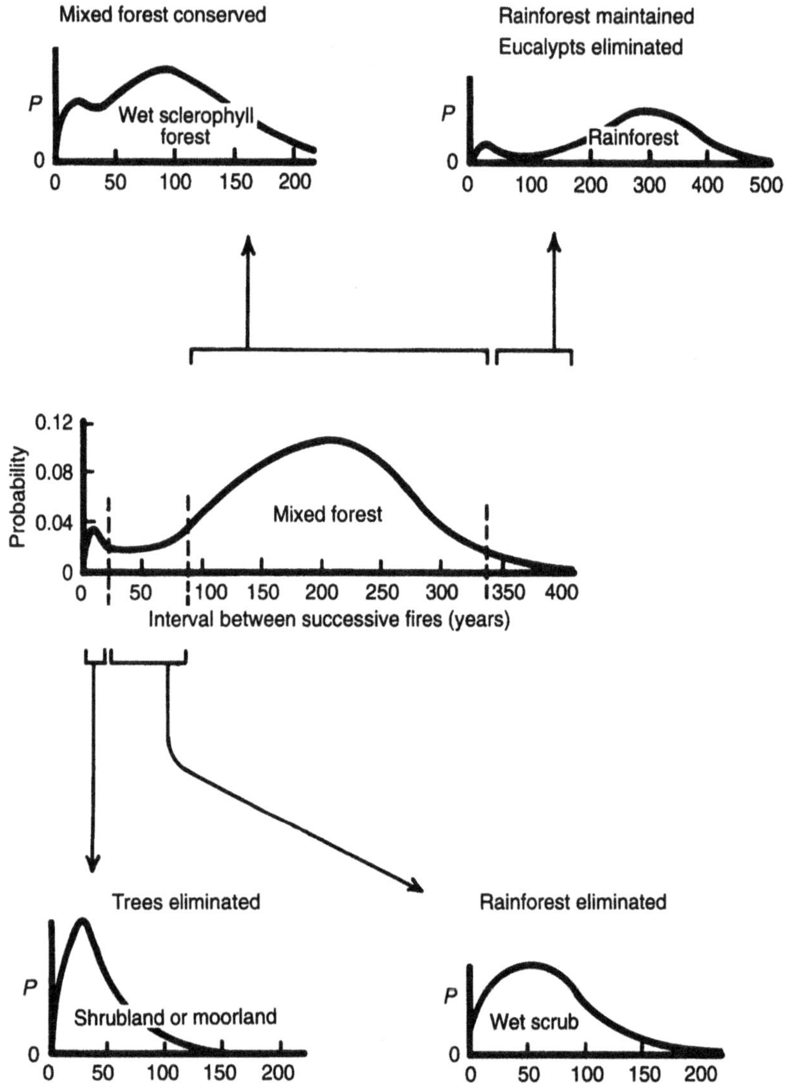

Figure 7.2 Successional pathways in Tasmanian forests in relation to different fire frequencies (from Jackson, 1968). The central diagram indicates the probability of fire as a function of stand age in mixed forests. The forests can be converted to shrublands by frequent burning which eliminates trees (bottom left) or wet scrub by less frequent burning which eliminates rainforest elements (bottom right). Once converted, the new vegetation develops a new fire probability curve. The upper part of the diagram indicates the effect of infrequent burning. Intervals longer than 300 years allow conversion to rainforest and elimination of eucalypts (upper right).

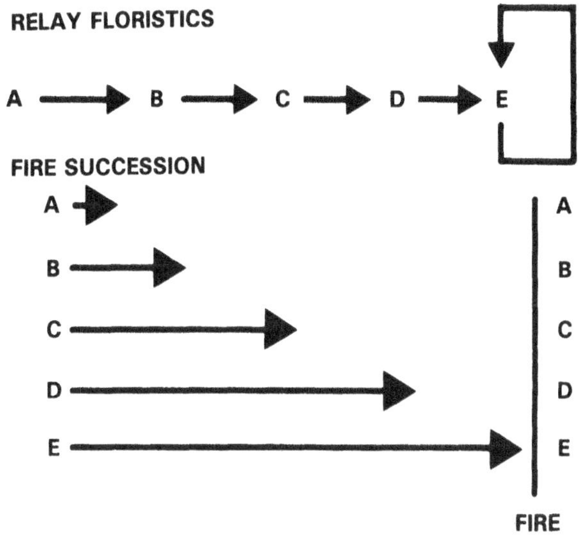

Figure 7.3 Diagram illustrating species replacement in old field succession (relay floristics) versus fire succession in shrublands (initial floristic composition). In relay floristics, competitive ability increases from left to right. In fire succession, longevity increases from left to right. Early successional species often persist in seedbanks. Note the absence of a self-replacing climax in fire succession.

Box 7.1 Life history attributes as determinants of succession after fire

Noble and Slatyer (1980) proposed a set of 'vital attributes' as the minimum information needed to predict successional change after fire. These included:

- mechanism of persistence during a disturbance, or arrival of propagules afterwards;
- ability to become established either immediately after a disturbance or later in the post-fire development;
- the timing of critical life-history events.

The potential methods of persistence during fire or arrival after fire were divided into the following categories:

- D (dispersed) species: arrival of widely dispersed seeds;
- S (soil-stored) species: seeds with a long viability, stored in the soil;
- G (germinating) species: as for S, but seed pools are exhausted after fire, and the species must reach maturity before the next fire to produce more seed in order to survive;
- C (canopy-stored) species: seeds with short viability surviving disturbance within protective fruits or cones;

(*continued*)

- V (vegetative) species: persistence by part of the individual surviving the disturbance, and recovery by vegetative growth, with immature tissue;
- U (unaffected) species: as for V species, but sprouting mature tissue;
- Δ species: a species with both D and U characteristics.

Similarly, three sets of species were recognized in respect of a species' ability to become established at some stage after fire:

- T (tolerant) species: able to establish at any time, tolerating competition with adults of the same or other species;
- I (intolerant) species: able to establish only immediately after disturbance, when competition is reduced;
- R (requiring) species: unable to establish immediately after a disturbance, but able to establish once mature individuals of the same or other species are present.

Thirdly, with reference to the times taken to reach certain stages in life history, the following were regarded as critical:

- the time taken for individuals to reach reproductive maturity (m);
- the life span of individual plants of the species (l);
- the time taken for all propagules to be lost from the community (e.g. for the seed store to become depleted) and for the species to become locally extinct (e).

Species types are described by a two-letter acronym in terms of this scheme. For example, *Cupressus forbesii*, a chaparral endemic, would be defined as a CI species since it is killed by fire, regenerates from seeds stored in the canopy (C) in serotinous cones and is intolerant (I), meaning that it establishes only immediately after fire. *Adenostoma fasciculatum*, chamise, would be classified as VI since it recovers vegetatively (V) from sprouting burls and has intolerant seedlings. A species designated DR, rare in fire-prone vegetation, would establish by means of propagules dispersed (D) in from outside the burnt area but would require (R) modification of the site by earlier successional species.

By incorporating the times taken to reach critical stages in life history, one can use species classifications based on these two vital attributes to predict successional response (see figure opposite). Critical stages are time to reach reproductive maturity (m), longevity (l) and seed longevity (e). Thus species relying wholly on soil seedbanks disturbance would become locally extinct if the fire interval exceeded $l + e$.

(*continued*)

Life-stage parameter characteristics for combinations of vital attributes ('species types') (after Noble and Slatyer, 1980). The critical events are the time to reach reproductive maturity, *m*, the longevity of the species population, *l*, and the longevity of its propagule pool, *e*. The (*m*) for some species types indicates the time taken to reach reproductive maturity when they invade a site not previously occupied by the species.

Species type	*Time since disturbance* 0---∞
DT, ST, VT	——*m*————————————————————————*le*
CT, GT	*m*———————————————————————————*le*
DI	——*m*———*l*—————————————————*e*
SI	*m*————*l*————————————*e*
GI	*m*————*l*—————*e*
CI	*m*——————*le*
VI	——*m*——————*le*
DR, SR	——*m*————————————————————————*le*
ΔT	*m*—(*m*)————————————————————*le*
ΔI	*m*—(*m*)———*l*——————————————*e*
UI	*m*————*le*

at a site (both established plants and their seeds) and their regeneration requirements in relation to disturbance. Community composition is altered primarily by changing the interval between fires (Chapter 4). The effect of altering disturbance frequency in woody communities would be due more to the timing of life-history events than to the elimination of superior competitors.

The difference from traditional views is clearest if one considers protocols for predicting successional dynamics in fire-prone communities. Instead of searching for competitive hierarchies based on relative resource use or growth rate (Keddy, 1989; Wedin and Tilman, 1993), a fire ecologist would begin by studying fire-survival response, reproductive timing and the life span of plants and their propagules. Competition hierarchies are largely irrelevant and would be added only if there were unexplained deviations from life-history predictions. Interactions between species are ignored as irrelevant for predicting successional change, or included as exceptions to the general rule.

The vital attributes approach has been most successful in predicting the likely outcome following fires in relatively species-poor woody communities (Noble and Slatyer, 1980). It is more cumbersome as the number of species increases or as more vital attributes are added. Recent versions have coupled vital attributes with individually based plant growth models

(Shugart, 1984) and include some consideration of competitive effects (Moore and Noble, 1990). The vital attributes approach is based on the assumption that interval-dependent processes (Chapter 4) dominate fire-prone communities. Event-dependent phenomena complicate the simple set of rules but, as examples in Chapter 5 indicate, are very important for predicting species response, especially where groups of species share the same vital attributes. In principle, they could be easily incorporated into a vital attributes model for predicting community dynamics (Table 5.8). The vital attributes approach is an essential starting point for most studies of community response rules to fire.

7.2.6 The importance of species interactions

The vital attributes approach can be viewed as a null model for predicting temporal change in plant communities because the dynamic behaviour of species depends on their intrinsic properties and not on interactions with other species. However, there are many fire-prone communities where species interactions may be important. The most well-studied departure from the non-interactive (null) view is, predictably, when species not only react to the prevailing fire regime but also alter it. The switch from flammable to non-flammable vegetation after long intervals without fire, as in the eucalypt example discussed earlier, is common in many regions with different floras and under different climatic conditions (Table 7.1). Many studies, in different parts of the world, have investigated the dynamic relationships between flammable communities and non-flammable, usually forest, vegetation (Mount, 1964; Jackson, 1968; Mutch, 1970; Kercher and Axelrod, 1984; Smith and Goodman, 1987; Archer *et al.*, 1988; Manders *et al.*, 1992). A typical example is the invasion of Cape fynbos by non-flammable evergreen forest species after long intervals without fire (Manders *et al.*, 1992). Early forest invaders are usually bird dispersed and perches are necessary for colonization (Archer *et al.*, 1988; Manders *et al.*, 1992). Juvenile plants are killed by fire but older plants resist by resprouting. As forest foci coalesce, the fuel properties change and regular fires are excluded, causing a complete system shift to one where species with reproductive dependence on fire are eliminated. Successional dynamics can be predicted from the biology of colonization, facilitation of recruitment by site alteration and the time required for trees to become resistant to fire, either by suppressing fuel or by developing the capacity to sprout. Relative growth rates or different patterns of resource use, emphasized in competition models (Grime, 1979; Tilman, 1982, 1988), would be superfluous for understanding the process.

The vital attributes approach has so dominated studies of fire succession that rather little is known about the alternative argument – that species interactions drive fire cycles. There are few explicit tests of whether the

Table 7.1 Examples of alternative stable states of frequently burned flammable and infrequently burned non-flammable vegetation. Most of the transitions between states have been shown by fire-exclusion experiments

Alternative states	*Region and sources*
Sclerophyll woodland → forest	Australia: Jackson (1968) N. America: Peet (1992)
Shrubland → forest	Africa: Luger and Moll (1992); Manders *et al.* (1992) Europe: Hobbs and Gimingham (1987)
Savanna → forest/thicket	Australia: Bowman *et al.* (1988); Gill *et al.* (1990) Africa: Trapnell (1959); Brookman-Amissah *et al.* (1980); Smith and Goodman (1987); Backeus (1992) N. America: A.S. White (1983) S. America: Coutinho (1982); San Jose and Farinas (1983)
Grassland → forest	Africa: Westfall *et al.* (1983) Asia: Stott (1988) N. America: Archer *et al.* (1988); Peet (1992)
Grassland → dune scrub	N. America: Peet (1992)

disappearance of species during a successional cycle is due only to finite life span or to suppression by other species. However, competitive suppression before a species can accumulate an adequate seedbank could ultimately eliminate a species from the successional community. Cowling and Gxaba (1990) studied the effect of a single species, a tall proteoid shrub, *Leucadendron laureolum*, on understorey composition in Cape fynbos. The shrub is a non-sprouter with densities that fluctuate widely within and between burns, depending on the accidents of wind dispersal and fire frequency and season. They found that where the shrub population was sparse, the understorey was rich in species, with a high proportion of ant-dispersed and non-sprouting species. However, in dense stands, diversity was half that of sparse stands, ant-dispersed species were rare and bird-dispersed shrubs became relatively more common, reflecting changing habitats for animals (Table 7.2). This study shows that community composition can be strongly influenced by interference from a single key species. However, these effects may be temporary. Community composition largely recovered after burning in this study, probably because suppressed species were able to persist

Table 7.2 The effect of different densities (stems m^{-2}) of *Leucadendron laureolum*, a tall proteoid shrub, on the species richness and proportion of species within functional types in the understorey of South African fynbos. Data are means from ten 1 m^2 quadrats for three homogeneous sites, each with paired sparse and dense proteoid stands in mature (15 yr) and recently burnt (3 yr) vegetation (from Cowling and Gxaba, 1990)

Understorey composition	*Mature stands*		*Recently burnt stands*	
	Sparse (0.1)	*Dense* (6.7)	*Sparse* (0.2)	*Dense* (3.5)
Species richness m^{-2}	13.1	5.7	20.6	11.5
% Ant dispersed	34.3	12.9	35.1	25
% Wind dispersed	22.9	16.1	20.2	21.1
% Bird dispersed	8.6	19.3	1.8	3.8
% Non-sprouters	60.6	32.2	71	61.5

in seedbanks until the next fire (Table 7.2). More studies of this kind are needed, especially in woody communities.

7.2.7 Grass–tree interactions: the Gulliver effect

In a few systems, but these of great extent, the dynamics of woody plants is largely determined by interactions with grasses. Here neither vital attributes nor competitive hierarchies provide much assistance in predicting dynamics. Their behaviour has been largely neglected in the theoretical literature but is of great applied interest. Grasses are extremely effective at suppressing regeneration of other growth forms because of their rapid growth rates (D'Antonio and Vitousek, 1992). Bunch grasses form highly flammable fuels but recover very quickly from burning. Since their growing point is near or beneath the soil surface and protected from all but extreme heat, fire destroys little more than a year's accumulated growth, most of which is dry and dead. Furthermore, grasses produce abundant seeds in 1 or 2 years following germination, whereas trees and shrubs take longer to mature.

The important interactions take place in the regeneration phase – producing what one may call the Gulliver effect (after Swift's famous sailor bound helpless by hordes of tiny Lilliputians). Gullivers are plants, typically stunted multi-stemmed shrubs, which dominate communities as adults but struggle to emerge from the herbaceous layer as juveniles. The herb layer interferes with Gulliver recruitment by suppressing seedlings, slowing the growth of established plants, and by fuelling frequent fires which kill or stunt the survivors so that they fail to escape the danger zone. Gullivers may persist for years surviving repeated burning by sprouting from ageing root systems. If Gullivers fail to emerge from this condition, they eventually die, though some species persist for decades in this condition (Menaut *et al.*, 1990). If the herb layer is removed, typically by grazing or burning, or if fires are suppressed, Gullivers are released from the danger zone and the vegetation can be rapidly transformed to a woodland. Thus Gullivers are hostage to the presence of small, weedy plants, low on Tilman's (1988) competitive hierarchy, but with fire on their side.

7.2.8 Savanna dynamics

The dynamics of savanna systems are notoriously complex. Vegetation change is contingent on combinations of drought, herbivores, humans, elephants, fire, insect outbreaks, competition and soil conditions, all interacting in complex and unpredictable ways (Norton Griffiths, 1979; Gillon, 1983; McNaughton, 1984; Trollope, 1984; Frost *et al.*, 1986; McNaughton *et al.*, 1988). It seems remarkable that such an unstable mix of grass and trees should exist, in dynamic equilibrium, over some 20% of the world's land surface. Explanations for the coexistence of grass/tree mixtures are diverse and controversial (Frost *et al.*, 1986; Belsky, 1990). One of the simplest and most general is based on apparent niche differences in grass/tree rooting systems (Walter, 1971). Trees can tap deeper subsoil sources of water than grasses, but grasses are more competitive in the soil surface layers. Walker and Noy-Meir (1982) used Lotka–Volterra type models to show that stable coexistence was possible with these differences in water use. They used their model to indicate how different soil types would affect tree:grass ratios because of differing water availability in surface and subsoil layers. For example, trees would tend to dominate deep, sandy soils at equilibrium because their deep roots would benefit from the deeper infiltration of water. Heavy grazing should also favour trees because, with less grass, more water would infiltrate to the deeper soil layers.

Neither the Walter hypothesis, nor Walker's formal development of it, take explicit account of the population biology of trees or fire. The hypothesis predicts a stable grass/tree balance unless the water balance is altered. However, many savanna studies have shown that fire exclusion alone can alter the grass/tree balance in favour of trees, contradicting niche-based arguments for coexistence (Chapters 4 and 5). Gulliver effects provide an alternative hypothesis for coexistence. Though no general mechanistic model for savannas seems to have been developed, key considerations have been quantified for a humid West African savanna (Menaut *et al.*, 1990) and for the Serengeti (Norton Griffiths, 1979; Pellew, 1983; Dublin *et al.*, 1990). Grasses inhibit recruitment of trees, primarily through burning. Trees, in turn, inhibit grasses and exclude fires once they become sufficiently dense (Figure 7.4). Gullivers can be released from the grass layer by heavy grazing or drought, both of which reduce the fuel load of grasses and prolong the fire return interval. The competitive effect of grasses on tree seedlings varies from system to system but is often less than their effect on fire, since Gullivers can escape even dense grass layers given sufficient time. Once the trees have grown above a threshold height (usually 2–3 m; Pellew, 1983; Trollope, 1984; Dublin *et al.*, 1990; Menaut *et al.*, 1990)

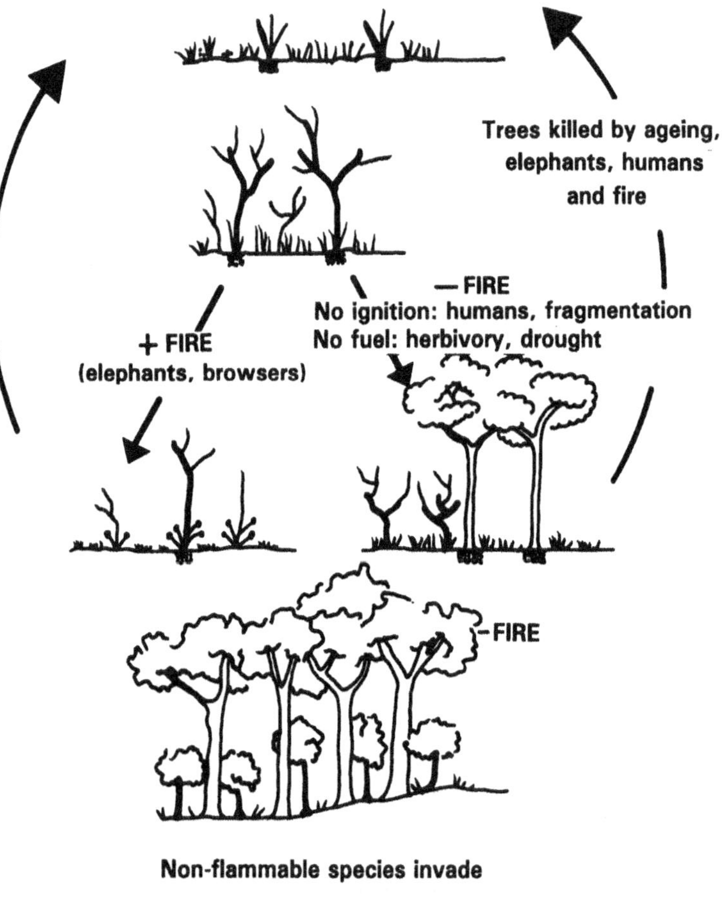

Figure 7.4 Elements of savanna dynamics. The top of the diagram illustrates tree seedlings and 'Gullivers', suppressed, multi-stemmed juvenile trees, in a grass layer. With frequent burning or browsing (centre left), 'Gullivers' are held in the grass layer. With infrequent burning due to low fuels or no ignition sources, 'Gullivers' periodically escape, forming savanna woodlands (centre right). With very long fire-free periods, non-flammable trees invade and fire is excluded (bottom).

burning no longer causes a regression to the grass layer and tree dominance increases (Figure 5.2).

There have been one or two attempts to model savanna as a fire-maintained equilibrium between trees and grasses. These have failed to show long-term coexistence between the two growth forms (Menaut *et al.*, 1990; Hochberg *et al.*, 1994). Trees usually end up dominating the community in model simulations because they are relatively immune from grass competition or fire once they have escaped the grass layer. A possible resolution to the problem is to identify additional processes that eliminate established trees. Natural thinning is one possibility. *Acacia mellifera* thickets became moribund and reverted to open savanna after 50 years of protection from grazing and fire in Namibia (Walter, 1971). Severe drought may cause periodic mass mortality in arid savannas, as may flooding in poorly drained soils. Fire can also occasionally kill large trees, especially where their bark has been removed by animal activity (Yeaton, 1988). Browsing mammals help keep Gullivers in the fire danger zone but seldom kill established trees. Elephants are a conspicuous exception. In the Serengeti-Mara ecosystem in East Africa, trees have steadily declined and been replaced by grasslands. Elephant damage to trees, even in the absence of fire, can prevent the return of woodlands (Dublin *et al.*, 1990). Finally, felling of trees by humans is a major factor in maintaining savannas, not least in Africa (Feely, 1980).

While all of these factors may contribute to mortality of trees, preventing savanna from becoming closed forests or thickets, they seem insufficiently general to explain the vast extent of this formation. For example, elephants cannot explain the persistence of the grass layer outside Africa. Part of the answer may lie in returning to the rooting niche arguments of Walter (1971) and Walker and Noy-Meir (1982). Savanna trees grow poorly if there is insufficient rooting depth, such as on waterlogged, sodic or iron-pan soils. Such soils, common in the catenary sequence of savanna landscapes, would act as refugia for grasses and foci from which fires would spread into adjacent woodlands. Trees, on the other hand, find refugia in rocky soils, on termitaria or other well-drained sites. Sustained coexistence of trees and grasses would then depend on the action of fire across a landscape matrix, with some facets favouring grasses and others favouring trees. New developments in modelling which incorporate spatially explicit information (e.g. Green, 1989; Menaut *et al.*, 1990; Hochberg *et al.*, 1994) may help resolve whether spatial heterogeneity is essential for the coexistence of trees and grasses in savannas.

In summary, fire is a highly selective mortality agent preferentially killing smaller plants. Competition or herbivory that retards growth greatly increase the hazard of subsequent fire death. Because of the unique properties of fire as a disturbance (large area, synchronized patch creation), very interesting dynamics emerge as a result of the disparities between the performance of juveniles and adults.

7.3 WHAT DETERMINES SPATIAL PATTERNS IN COMMUNITIES?

At least two rather different sets of response rules seem appropriate for fire-prone vegetation – the vital attributes approach for woody communities and grass/tree competitive and suppressive effects in savannas. Spatial pattern in fire-prone vegetation is strongly linked to successional dynamics at a variety of scales. To what extent are spatial patterns driven by the same kinds of considerations?

There are two distinct causes of landscape pattern:

- environmental variation related to soils, topography, climate;
- stages of recovery after disturbance in the same environment (Plate 3).

We first consider patterns along environmental gradients. There are several descriptive accounts of species turnover along environmental gradients (beta diversity), but explanations for spatial patterns are still sparse. The classic assumption has been that species are found where their physiologies best fit the physical environment – their fundamental niche. Experimental studies have shown that this is often not the case (e.g. Keddy, 1989; Bertness, 1991). The realized niche in these studies is smaller than the physiological capabilities of the plants. Interactions, especially interspecific competition but also facilitation, may therefore account for the position of species along resource gradients.

What causes spatial patterning in fire-prone communities? If competition is unimportant in driving fire succession, is it also unimportant in structuring spatial patterns? Does spatial variation in fire regimes cause spatial gradients in plant communities? These questions are relatively unexplored in fire ecology and there is much scope for further study and new insights. We first discuss evidence for fine-scale patterning in communities as a result of patchy fire behaviour. We then turn to landscape-scale studies of how fire generates vegetation patterns due to the complex interplay between fire regime, fuels and local weather patterns.

7.3.1 Fine-scale patterning

Fire behaviour is affected by many factors, including weather, direction of burn and topography, but also the nature of the community. Fire behaviour is especially variable in woody communities, where the size of the plants produces heterogeneity in the horizontal and vertical distribution of fuels (DeBano *et al.*, 1979; Hobbs and Atkins, 1988; Figure 7.5). It is tempting to speculate that individual species might create the fuels and local fire conditions most suitable for their own propagation. However, no such feedback effects have been reported to our knowledge. Nevertheless there is growing evidence that local differences in fire behaviour influence vegetation patchiness. Ash bed effects are extreme examples caused by the intense heat

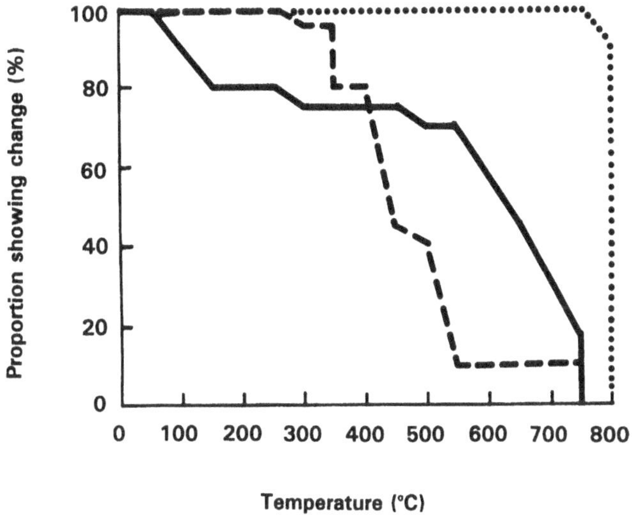

Figure 7.5 Spatial heterogeneity of soil-surface temperatures in three experimental fires in south-western Australia (from Hobbs and Atkins, 1988). Temperatures were recorded using heat-sensitive paints at 10–20 localities in a 20 × 20 m area in each fire. The graph indicates the proportion of localities reaching a particular temperature.

generated when a log or other fuel concentration burns. This can result in local nutrient enrichment, but can also affect recruitment by killing seeds or stimulating germination of hard-seeded species (Vines, 1980). Less extreme local variation in fire intensity also influences community patterns; for example, in chamise-chaparral. Rice (1993), using an ordination technique, found a correlation between post-fire vegetation establishment and fire intensity at a scale of metres. The correlation was attributed to phosphate fertilization associated with ash deposition. An alternative explanation is that local variation in soil temperature influenced seed survival and germination (Chapter 5). It is interesting to note that hard-seeded legume species (*Lotus*) thrived in the hot spots in Rice's study and are stimulated to germinate by intense heat (Moreno and Oechel, 1991a).

Ne'eman *et al.* (1992) reported similar patterns in a burnt Mediterranean *Pinus halepensis* forest. The fire burnt intensely under the pines. Post-burn recruitment of both herb and shrub species was lowest near burnt skeletons and highest away from the pines. A single species, *Rhus coriaria*, had more seedlings under the pines. The scarcity of seedlings around each skeleton was thought to be due to high seed mortality in the pine 'hot spots' or to the inhibitory effects of ash on seed germination. Neither of these studies explored pre-burn heterogeneity in seedbanks, which may have depended on the pre-burn overstorey. Nevertheless, it seems highly likely that local variation in fuels, leading to patchy fire intensity and hence patchy seed survival and germination, contributes to small-scale community pattern.

A different kind of fine-scale patterning occurs in *Banksia* recruitment in Western Australian shrublands (Enright and Lamont, 1989; Lamont *et al.*, 1993). Here the post-fire environment is a mosaic of open, sandy patches interspersed with patches of incompletely burnt surface litter. The seeds of serotinous species aggregate in the litter patches, along with other debris distributed by wind and water. Seedlings undergo intense density-dependent thinning in the litter sites but not in the open sandy patches where they are too sparse. The relative proportions of litter and open sites can be altered by fire intensity changing the distribution of seeds, strength of seedling interactions and the population sizes and spatial distribution of serotinous shrubs (Table 7.3). This local post-burn patterning is probably quite general in guilds of serotinous species.

These examples show that fire regime influences the patchiness (heterogeneity) of plant communities. The effects of fire on heterogeneity at this scale are still poorly understood but may have important implications for species coexistence and the maintenance of species richness in plant communities (see below). Patchy nutrient distribution has been invoked to explain coexistence in competitively structured plant communities (Tilman, 1982). Patchy fire intensity may have analogous effects in fire-prone communities. The number of species coexisting in a community may then depend on spatial heterogeneity of fuels.

7.3.2 Coarse-scale patterning

Fire may pattern communities at far larger scales. Beta diversity is usually attributed to changing species performance across resource gradients, but consistent differences in fire behaviour also cause community turnover. Fire

Table 7.3 Effects of different fires on post-burn patchiness and *Banksia* seedling recruitment in Western Australia (from Enright and Lamont, 1989). Post-burn sites were either open and sandy or covered with litter. Wind-blown *Banksia* seeds accumulated in the litter site. Data derived from a sample of 285 seedlings in the spring burn and 684 seedlings in the autumn burn, collected in 5 × 25 m^2 quadrats in each burn

	Spring burn		*Autumn burn*	
	Sandy sites	*Litter*	*Sandy sites*	*Litter*
Patch area (%)	70	30	86	14
Seedlings m^{-2}				
At start of summer	0.7	5.7	2.5	23.2
At end of summer	0.4	4.1	1.3	5.5
% Mortality	42	28	48	76

behaviour changes across local edaphic and topographic gradients because of differences in fuel moisture content, fuel properties and the effects of topography on wind (Wells, 1965; Cheney, 1981). The most obvious result is a patterning of the landscape into more and less flammable communities (Plates 4–6). In grassland landscapes, shrubs and trees are generally confined to rocky scarps which act as refugia where fire intensity is reduced by the discontinuous fuel layer (Wells, 1965, 1970). Abrupt forest/prairie boundaries in Minnesota coincide with large rivers and lakes acting as natural firebreaks and not with climate or soil differences (Grimm, 1984). Islands of mesic non-flammable forests in a sea of flammable vegetation are common in ravines, and on rocky aprons or river banks which afford some protection from burning (Ash, 1988; Geldenhuys, 1994). Figure 7.6 illustrates the landscape pattern observed for patches of evergreen forest surrounded by fynbos along the southern coast of Africa (Geldenhuys, 1994). These forests have abrupt (1–20 m) margins with adjacent flammable communities but the margins do not coincide with soil or geological changes. Geldenhuys (1994) has shown that the distribution of forests in this landscape can be explained by fire movement patterns through the adjacent fynbos driven by extremely hot 'bergwind' conditions. Fire suppression, in this and other instances, has allowed the spread of the non-flammable species into the surrounding vegetation, indicating that their physiological tolerance is broader than their site occupancy in the presence of fire (e.g. Luger and Moll, 1992; Figure 7.7).

It is also interesting to ask whether consistent differences in fire regimes can generate species or community turnover *within* fire-prone vegetation. This could happen if species differed in fire responses, producing a sorting of species along gradients of fire frequency or intensity. The fire-response component of the fundamental niche has seldom been considered in studies of distribution limits. Grimm (1984) showed that gradients in tree composition along the east margins of the Big Woods area of Minnesota coincided with gradients in fire frequency, from regularly burnt prairies to virtually fire-proof deciduous forests. The sequence of tree species along the gradient was consistent with fire tolerance rankings, with aspens and oaks on prairie margins and fire-sensitive sugar maple in the least fire-prone forests. Here a similar ranking of shade tolerance may fool one into believing in a competitively driven gradient when fire is the simpler explanation (Grimm, 1984).

Bergeron and co-workers have provided several interesting examples from fire-prone Canadian boreal forests (Bergeron and Brisson, 1990; Bergeron, 1991), suggesting that fire regime, and not climate, determined northern distribution limits of species. The fire regime of boreal forests is dominated by crown or surface fires of high intensity and very large area. These favour serotinous conifers (*Picea mariana*, *Pinus banksiana*; Chapter 6) or vigorously sprouting birch and aspen. Red pine (*P. resinosa*) is a non-serotinous species with a different fire-response niche. It is killed by crown

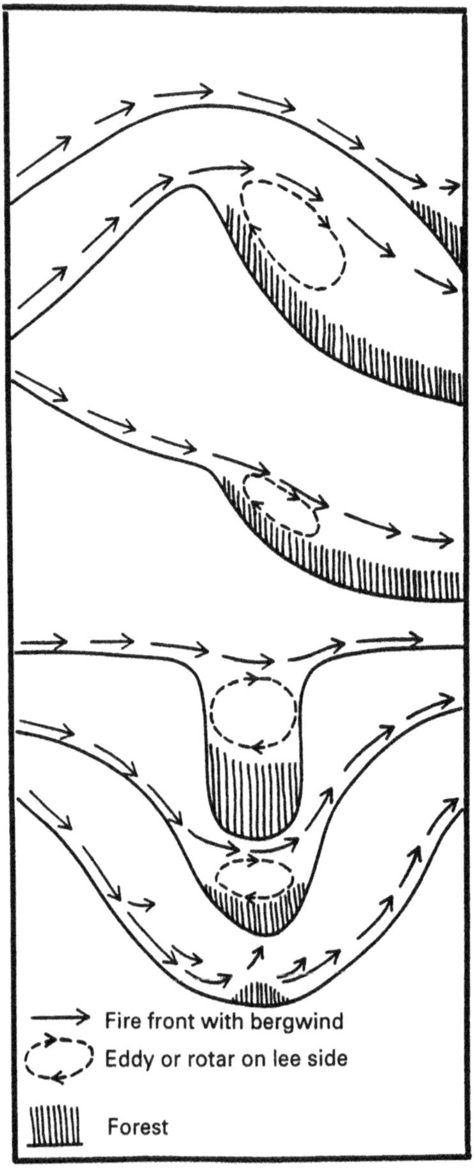

Figure 7.6 Schematic view of the relationship between wind movement, topography and forest distribution in the southern Cape, South Africa (Reproduced from Geldenhuys, Blackwell Scientific Publications, 1994). Intense fires develop in fynbos under hot, dry 'bergwind' conditions. Wet evergreen forests persist in wind shadows in the lee of topographic barriers where they are less prone to burning.

fires, but tolerant of surface fires of light to moderate intensity. In the northern parts of its range, red pine is restricted to lake landscapes and to islands. Extensive evidence from fire scars, tree rings and size analyses showed that fires over the past two centuries occurred at higher frequencies

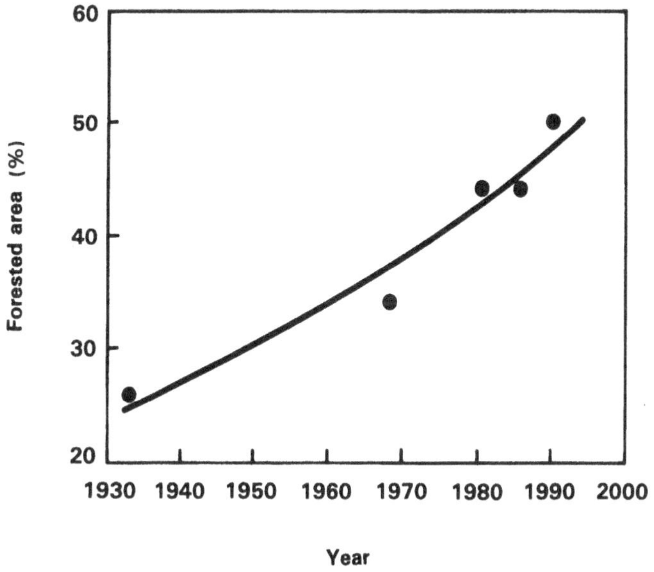

Figure 7.7 Invasion of South African fynbos shrublands by evergreen forest over a 58 year period without fire (from Luger and Moll, 1992). The forested area (percentage of total area of 2500 ha) was determined from a sequence of aerial photographs supplemented with ground observations.

but lower intensities on islands than on the adjacent mainland (Bergeron and Brisson, 1990; Bergeron, 1991). As a result, red pine finds refuge on the islands, while the mainlands are dominated by the serotinous conifers. The disturbance regime, and not physiological tolerance, explained the northern limits of the species (Plate 4). Studies such as these, exploring new scales of time and space, offer exciting prospects and novel explanations for the old ecological problem of distribution limits. They need to be extended to other fire-prone systems.

7.3.3 Patterns resulting from different stages of recovery

Pattern generated by different stages of post-fire recovery is obvious in fire-prone landscapes and little subtlety is needed in identifying this cause of spatial patterning (Plate 3). The mosaic of communities in different stages of post-fire succession has significant influences on wildlife habitat, stream flow, nutrient cycling and other ecological processes. Some of the best studies of landscape pattern created by disturbance come from North American conifer forests. For example, Romme (1982) studied the effect of fire suppression on vegetation diversity at the landscape level in the subalpine forests of Yellowstone Park. The area has a relatively uniform topography and geology but is separated into two major habitat types on the basis of rainfall differences. Past fire history is responsible for community variation within these. Five successional communities can be distin-

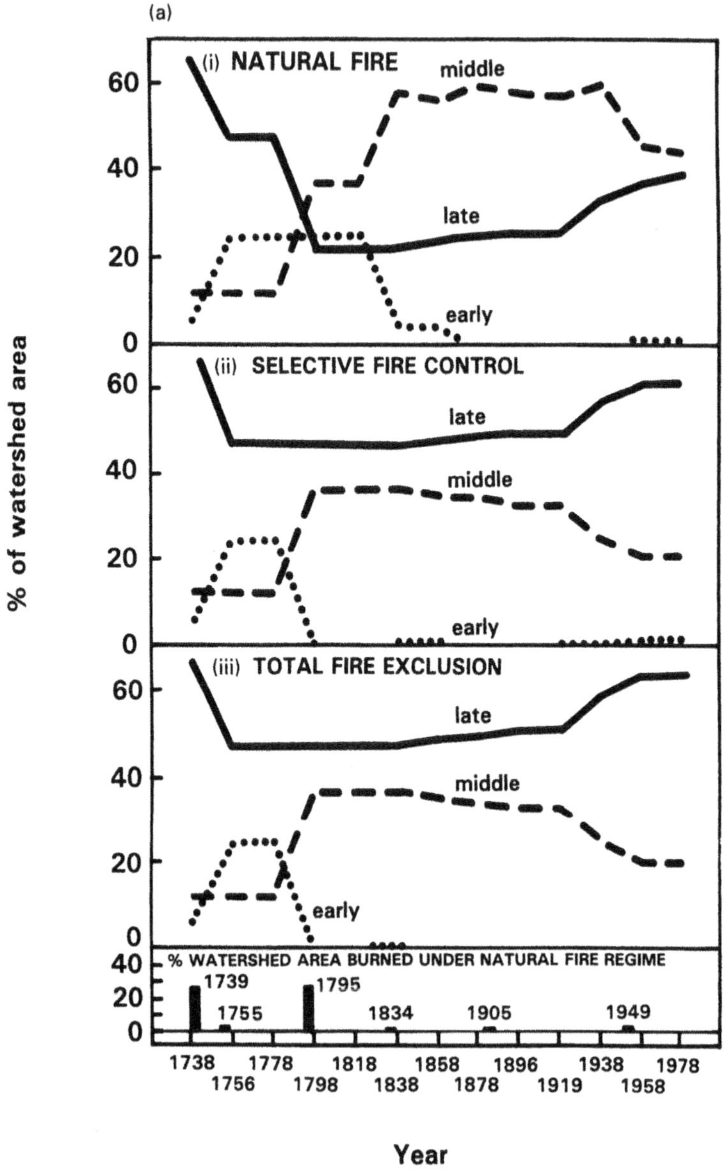

Figure 7.8 The effects of fire on landscape diversity in subalpine forests of Yellowstone National Park (from Romme, 1982). Early, middle and later successional stages were recognized based on age since burn. The sequence of stand-age mosaics was reconstructed from fire-history data. Landscape reconstructions were made using a simulation model based on hypothetical fire management policies. (a) Percentage of a watershed area covered by early, middle and late successional stages for the three fire policies: (i) natural fire – all fires allowed to burn; (ii) selective fire – large fires suppressed; and (iii) total fire exclusion. (b) The derived landscape diversity statistics expressed relative to their maximum possible value.

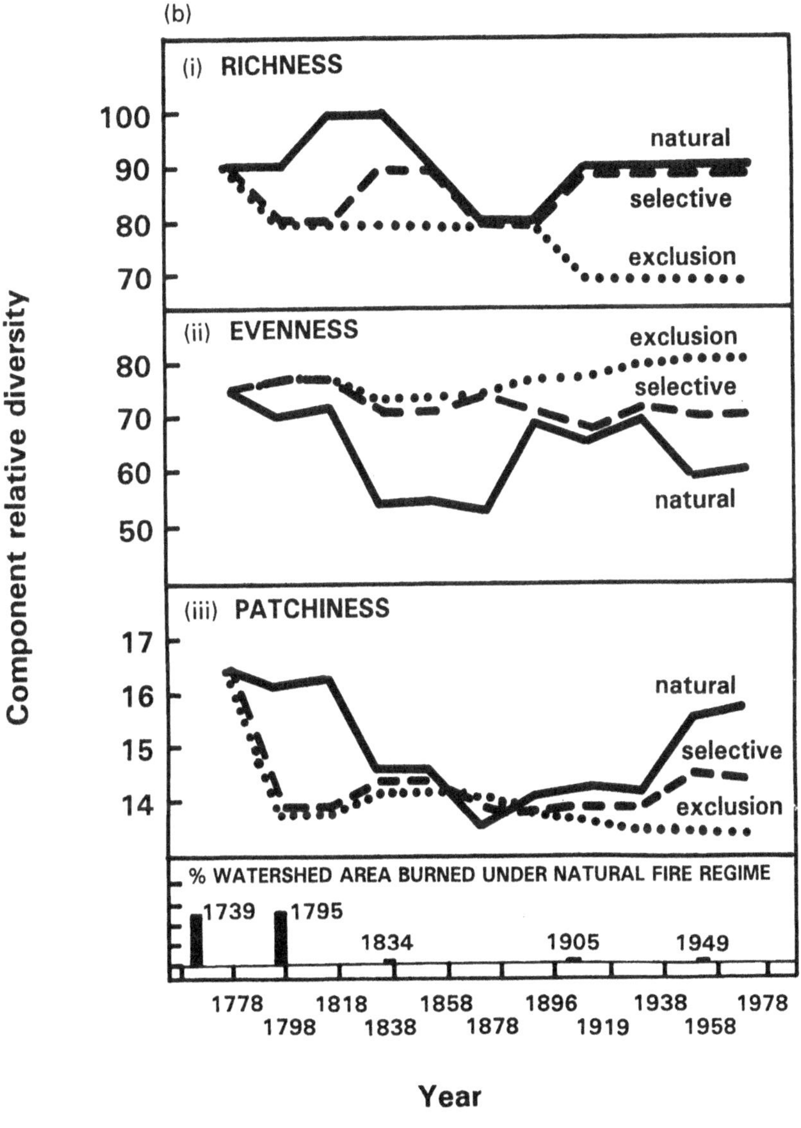

guished in each type based on interval since the last fire. Fire-history data for the past 200 years, based on analysis of fire scars, were used to reconstruct the changing vegetation mosaic of these successional communities. From the historical records, changes in landscape diversity could be assessed from the number of community types and their evenness (the proportion of the landscape covered). Landscape diversity depended strongly on fire history and, to a lesser extent, mountain pine beetle depredations. Patterns were highly variable and it is clear that there is no 'equilibrial' distribution of seral stages, even though the area is large (73 km^2). Fire suppression or selective fire control (permitting only small fires to burn) would have had significant effects on landscape diversity, had they been applied for the past 200 years (Figure 7.8a).

Whether such effects would be 'good' or 'bad' is difficult to say. Clearly there is no 'balance of nature' at Yellowstone in the sense of equilibrial proportions of different aged stands (Sprugel, 1991). Criteria for setting fire management objectives must be very difficult.

7.4 WHICH COMBINATIONS OF SPECIES OCCUR TOGETHER AND WHY?

Though the above discussion indicates a rich literature on community response rules, far less is known about assembly rules in fire-prone communities. Here the problem is not how and why species assemblages change with time but which species make up an assemblage and why. Why are some combinations of savanna trees common and others not? Why is chaparral dominated by one or two shrub species while fynbos is dominated by many? We have suggested that competitive interactions may be irrelevant for predicting successional changes within many fire-prone communities. This does not mean that they are unimportant in determining species turnover along resource gradients or in regulating the number of species that coexist within a fire successional niche.

The question of what determines coexistence and species richness at a local scale (alpha diversity) has been a major focus of plant community studies elsewhere. The competitive exclusion principle argues that, at any one site, the species better able to grow and reproduce should exclude all lesser competitors so that species richness should be uniformly low (Crawley, 1986). Since this is almost universally not the case, a plethora of explanations for departure from the principle have been put forward (Crawley, 1986; Keddy, 1989). The problem has been relatively neglected in fire-prone communities although there are excellent opportunities for experimental testing of some of the current hypotheses. We briefly discuss some alternatives and relevant studies which illustrate the possibilities.

7.4.1 Competition and coexistence in pyrophytic shrublands

Coexistence studies have been neglected in fire-prone systems, perhaps because of the belief that fire so disrupts the equilibrial state that species never interact sufficiently strongly for competition to proceed to exclusion. However, we note from Chapter 4 (Figure 4.3) that even with extreme stochastic fluctuation in population size, populations will generally grow exponentially until they 'explode'. Some form of intra- or interspecific containment must be operating. The regular occurrence of burning does not explain away the problem of how so many species coexist in the same community.

There has been a spate of recent studies on coexistence in Californian chaparral and its southern counterparts in Australia and South Africa

Table 7.4 Hypotheses for coexistence in plant communities (modified from Crawley, 1986)

Hypothesis	*Description*
1. Refuges (successional niches)	Inferior competitors may persist in a community if there are refuges from competitors (e.g. seedbanks)
2. Niche separation, resource partitioning	Coexisting species use resources differently so that competitive exclusion does not occur
3. Recruitment lotteries	Site occupancy is founder controlled; no competitive exclusion; random variation in seed production permits coexistence
4. Dispersal differences	Inferior competitors are superior dispersers and avoid exclusion through dispersal
5. Spatial and temporal heterogeneity	Species coexist because stochastic variation never consistently favours one species, especially in the regeneration niche
6. Disturbance	Disturbance prevents competitive exclusion
7. Selective herbivory	Dominant competitors are the preferred food of herbivores, which indirectly allow coexistence of weaker competitors

which have explored a number of the hypotheses listed in Table 7.4 (Davis, 1991; Bond *et al.*, 1992). Perhaps the most widely held view is that species segregate temporally, occupying different periods in the post-fire successional cycle. Coexistence of all species in the cycle can thus be readily explained by different 'successional' niches. Short-lived species which are 'lost' during post-fire succession, mostly persist in refugia (seedbanks) through the long fire-free interval (Table 7.4, hypothesis 1; Chapters 3 and 4). Though interactions among species in different successional niches may occur, they would have negligible effects on community composition in the next fire cycle where these seedbank refugia exist.

Successional niche differentiation does not explain the coexistence of species with the same successional niche and with similar growth forms. Members of the Proteaceae often dominate Australian and South African shrublands. Many of the species have similar growth forms, growth rates and life spans yet coexist in mixed stands. There is growing interest in mechanisms that allow coexistence of these fire-dependent shrubs. Cody (1986) explored distribution patterns of proteoid shrubs in the species-rich Cape fynbos shrublands. He asked whether co-occurring species were a random selection from the regional species pool or whether they occupied different niches as measured by differentiation on leaf morphology axes (Table 7.4, hypothesis 2). He found no overlap in leaf shape and size in coexisting species, suggesting that this guild is, indeed, segregated according to some resource axis represented by leaf morphology (Figure 7.9). In several cases the same species had different leaf morphologies in different localities but always maintained non-overlapping niche space relative to congeners.

Cody's study is interesting in suggesting that competition may structure shrub communities, in contrast to the traditional emphasis of fire ecologists on the fire regime and physical site factors as determinants of distribution patterns (Chapters 4 and 5). Similar evidence for niche differences has been observed in chaparral, where both shoot and root architecture differ in coexisting shrubs (Davis and Mooney, 1986; Davis, 1991). There are several difficulties with interpreting these data as evidence for niche structuring of community composition. Cody suggested that the different leaf forms were an expression of different patterns of water use. Davis, however, noted that even though clear differences in water use existed amongst chaparral shrubs, there was no direct evidence that these were sufficient to allow coexistence. It is not at all obvious just how different niches should be to allow coexistence. The problem may be complicated further if coexistence is determined by interactions at the regeneration stage. Lamont and Bergl (1991), working on three co-occurring Australian *Banksia* species, very similar to fynbos proteoids, measured differences in water use of mature plants directly. They found no differences in transpiration rates or leaf water potentials among the species, all with very similar leaf morphologies, and argued that reproductive niche differences were a more likely explanation for coexistence.

Niche differences have been identified at seedling stages and not only among mature plants (Lamont *et al.*, 1989; Davis, 1991). Lamont *et al.* (1989) tested niche differentiation in the regeneration phase among five *Banksia* species in Western Australian shrublands. These species show a replacement sequence along an elevational gradient. *Banksia hookeriana* is a species entirely restricted to the mid-elevation site. Seeds of this species were planted at three points on the gradient in monocultures and in mixtures with its congeners. Germination rates were comparable across sites, as was seedling survival in monocultures (Table 7.5). However, in mixtures, *B. hookeriana* survived only in mid-elevation sites, its natural habitat. There were marked differences in water status of seedlings grown in mixtures and monocultures at two of the sites but not at *B. hookeriana*'s natural distribution centre. This is an interesting example of apparent competitive sorting of species very early (7 months) after establishment. If it proves more general, seedling attributes will need much closer scrutiny than has been the case in the past.

Lottery models are a novel non-equilibrial explanation for coexistence of competitively similar shrubs based on regeneration differences (Chesson and Warner, 1981; Fagerstrom, 1988; Table 7.4, hypothesis 3). Lotteries assume founder control – that competition among recruits is for space – and

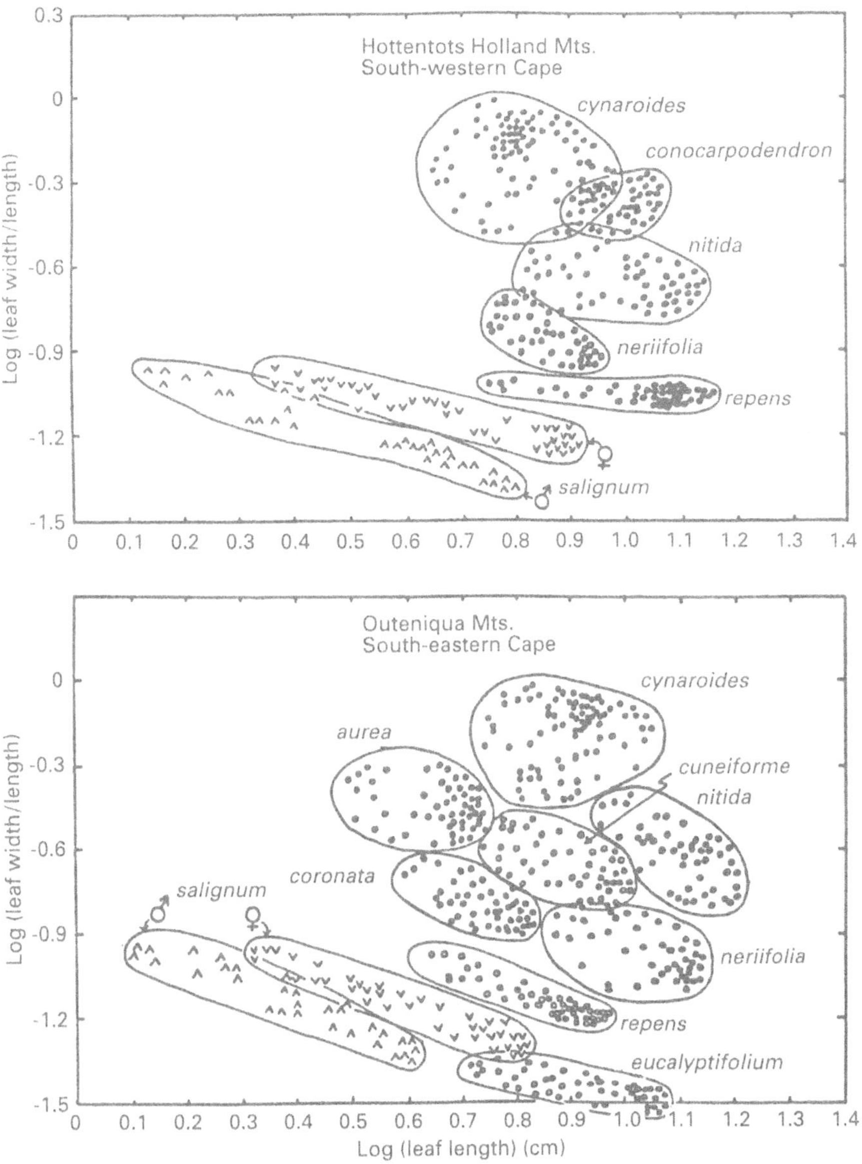

Figure 7.9 Leaf morphology variation in co-occurring shrub members of South African Proteaceae (from Cody, 1986). The lack of overlap between species clusters in this 'leaf morphology plane' was interpreted by Cody as evidence for niche differentiation in the proteoid guild. (From Figure 23.9 (B) & (C) from *Community Ecology* by Jared Diamond and Ted J. Case. © 1986 by Harper & Row Publishers, Inc. Reprinted by permission of Harper Collins Publishers, Inc.)

Table 7.5 Recruitment of *Banksia hookeriana* in monocultures and mixtures with other *Banksia* species at three positions on a topographic gradient in south-western Australia (from Lamont *et al.*, 1989). Germination and survival data (1 year after sowing) are given as a percentage of initial seeds planted after a fire

Life-history stage		*Lower slope*	*Mid-slope (preferred site)*	*Upper slope*
% Germination	mixed	55.0	70.0	62.5
	mono	70.0	57.5	67.5
% Seedling survival	mixed	0.0	32.5	0.0
	mono	37.5	37.5	47.5
Adult stems ha^{-1}		0.0	670	0.0

that space is allocated randomly (as a lottery) to propagules of the different species. The winners of a lottery will be in proportion to the reproductive success of competing species. For long-term coexistence, species should vary in reproductive success so that one species is not *consistently* the stronger recruiter. The effects of a good recruitment year for one species can be 'stored' over the lean periods when other species are favoured. For example, a resprouting species that established well after one good recruitment episode could persist through many fire cycles with poor recruitment until the next good burn. The storage effect is not restricted to sprouting species. Lottery models are just as applicable to non-sprouters (Bond *et al.*, 1992).

Lottery models may have rather limited application, in their pure form, to problems of plant coexistence. In the post-fire environment, more seedlings will generally arrive at a gap than the space available for an adult. Since fire synchronizes recruitment, most seedlings start growing within a few months of each other. Lottery models imply scramble competition where all plants suffer equally from crowding so that species with the most seedlings win the lottery. However, contest competition is common in plants where small differences in species growth rates translate to large differences in competitive outcome (Crawley, 1990; Weiner, 1990). Situations where seedlings are so sparse that they do not compete for the same space, or where species growth rates are exactly matched and neither species is a consistent winner, must be very rare. Australian and South African proteoids may be one such case. One testable assumption of lotteries is that the relative abundance of competing adult populations should be proportional to their relative abundance as seedlings, but the experiments have yet to be done.

Differences in dispersal provide yet another explanation for coexistence if the poorer competitor is also the better disperser (Table 7.4, hypothesis 4). Yeaton and Bond (1991) tested a possible example in two Cape proteoids. One species, *Protea lepidocarpodendron*, has serotinous cones with seeds dispersed by wind after fire. The morphologically similar *Leucospermum conocarpodendron* is dispersed by ants. The wind-dispersed species is the superior competitor, soon overtopping the ant-dispersed species and severely reducing its growth, reproduction and fire survival (Table 7.6). This is an example of the problem with pure lottery competition. A single *Protea* seedling would win occupancy of a site regardless of the number of *Leucospermum* seedlings competing for the same site. Nevertheless, differences in dispersal fates of the two species may allow coexistence. Wind-dispersed seeds accumulate under the skeletons of burnt shrubs after fire, including those of the weaker ant-dispersed competitor. However, ants carry away the seeds from the shade of these tall shrubs and into nest entrances in openings where the wind-blown *Protea* seeds are least likely to accumulate after fire (Table 7.7).

Yeaton and Bond used Markov simulations to predict whether stable

Table 7.6 The effects of *Protea lepidocarpodendron* on *Leucospermum conocarpodendron* in Cape fynbos. Both species are 2–3 m tall shrubs but the *Protea* is wind dispersed and the *Leucospermum* ant dispersed (Yeaton and Bond, 1991). Values are means (SE in brackets)

Measure of Leucospermum *performance*	*In the open*	*Next to* Protea
Stem diameter (cm) of 15-year-old shrubs	6.9 (0.2)	3.3 (0.2)
n	25	25
Number of inflorescences/shrub	8.9 (1.7)	0.2 (0.2)
n	25	25
% Surviving fire ($n = 100$)	56.5	21.1

Table 7.7 The density of ant-dispersed *Leucospermum conocarpodendron* seedlings in three post-burn microsites relative to their wind-dispersed competitor, *Protea lepidocarpodendron* (Yeaton and Bond, 1991). L:P is the ratio of *Leucospermum* to *Protea* seedlings in each site. Values are mean number of seedlings m^{-2} (SE in brackets) from fifty 1 m^2 quadrats in each patch type

Seedlings	*Post-burn microsite*		
	Open	*Under burnt* Leucospermum	*Under burnt* Protea
Leucospermum	1.7 (0.4)	2.6 (0.3)	0.6 (0.1)
Protea	0.3 (0.1)	3.5 (0.7)	12.8 (7.1)
L:P	5.7	0.7	0.05

coexistence was possible with these dispersal differences. In this case, dispersal differences were not sufficient to promote coexistence. The more competitive *Protea* populations were predicted to increase rapidly, eliminating open space, and excluding the *Leucospermum* after only two or three fires. Only periodic fires which selectively eliminated the stronger competitor but not the weakly competitive *Leucospermum* species could allow temporary coexistence. Mixtures of the two species are rare in nature, so the distribution of the *Leucospermum* appears to depend partly on fire history but mostly on surviving in sites avoided by *Protea* populations.

Spatial or temporal heterogeneity in fire regimes has also been suggested as a potentially important explanation of coexistence in founder-controlled communities (Cowling, 1987; Table 7.4, hypothesis 5). Spatial variation in fire intensity may eliminate some species from a patch while favouring others (Moreno and Oechel, 1991a; Rice, 1993). Temporal variation in fire frequency, as in Yeaton and Bond's study, would have similar effects. This hypothesis is attractive because spatial and temporal heterogeneity are such obvious features of fire, especially in heterogeneous woody fuels. In communities dominated by fire-recruiting non-sprouters, plants only have opportunities for recruitment immediately after fire. Patchiness in seed survival would translate to patchiness in species distribution persisting until burning stimulates the next bout of recruitment. Crawley (1986) has suggested that the heterogeneity hypothesis is flawed because by explaining everything it explains nothing. However, there are testable predictions, at least in principle. For example, species richness should decline with decreasing variance in the fire regime. Fires at fixed intervals should result in loss of species relative to more varied intervals (Cowling, 1987). This prediction has important implications for managing fire for conserving diversity but has yet to be tested.

The problem of proteoid coexistence is far from resolved. However, these studies do show that the problem does not disappear because communities are frequently disturbed. Interactive processes are interesting and important, especially for species occupying the same successional niche in the fire cycle. Community assembly rules, even in highly dynamic fire-dependent guilds, must take cognizance of which species occur together.

7.4.2 Disturbance-mediated coexistence

The competitive exclusion principle applies to species coexisting at equilibrium. Disturbance may prevent competition from proceeding to exclusion thus allowing non-equilibrial coexistence (Table 7.4, hypothesis 6). Disturbance-mediated coexistence has been well studied in marine and some terrestrial systems and could be expected in pyrophytic vegetation. Coexistence of trees and grass in savannas could be viewed as an example where disturbance allows the persistence of unequal competitors because of the Gulliver effects discussed earlier. Examples appear to be rare of the more conventional case of disturbance-mediated coexistence between a dominant and its weaker competitors within the same growth form (e.g. Yeaton and Bond, 1991), perhaps because fire ecologists have seldom studied competition in conjunction with disturbance response.

7.5 DETERMINANTS OF SPECIES DIVERSITY

7.5.1 The intermediate disturbance hypothesis

If disturbance mediates coexistence of species, it may also explain species richness of whole communities. Very high rates of disturbance might eliminate all but a handful of species tolerant of these extremes. Low rates of disturbance would provide sufficient time for competitive exclusion and also lead to species loss. Diversity would be at a maximum at some intermediate frequency of disturbance which prevents competitive exclusion (Connell, 1978; Grime, 1979; Huston, 1979). Much of the experimental support for the intermediate disturbance hypothesis comes from sessile marine communities which have the advantage of responding very rapidly to experimental manipulation (Sousa, 1984). Is species richness highest at intermediate fire frequencies?

Within a single cycle of post-fire succession, species richness typically peaks in the first year after fire and then declines as the community ages (we will ignore those cases where tree species invade and convert the community to a less flammable state, though noting the problem this poses). As we noted earlier, this decline is usually attributed to the loss of species with short life spans and not to competitive exclusion, at least in woody communities. Would systems driven by vital attributes, rather than by competitive hierarchies, also show peak diversity at intermediate levels? It seems plausible. Very short fire return intervals would eliminate non-sprouting species with long juvenile periods. Very long intervals would exceed the life span of some species and especially their seedbanks. Peak diversity should occur at intermediate fire frequencies. Both mechanisms might operate simultaneously in some communities.

Tests of the intermediate disturbance hypothesis, which relates to the effects of different fire return intervals, have been confounded with diversity patterns within a single interval. The number of species generally declines with time since burn, but many species persist in dormant seedbanks. Ideally, species richness should be censused at its peak early in the fire cycle so that comparable stages of fire succession are compared. One of the few studies where this was done was that of Hobbs *et al.* (1984b) who counted the number of species in a chronosequence of Scottish heath in the first year after fire and also estimated species persisting in the seedbank at different stand ages (Figure 7.10). Their results showed the expected decline in species after long intervals without fire. Stands burnt at intermediate intervals (6–15 years) had the highest diversity. There were no

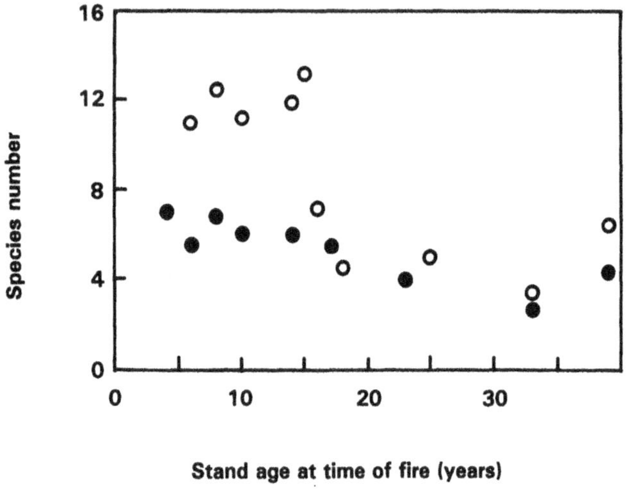

Figure 7.10 Relationship between vascular plant species richness one growing season after fire (open circles) and stand age when burnt in Scottish heath (data from Hobbs *et al.*, 1984b). Data are means from four 1 m^2 quadrats from each stand age. The number of species germinating from soil samples taken independently in stands of different post-burn ages is also shown (closed circles, mean of five samples per stand).

stands burnt at intervals shorter than 6 years, so the predictions of the intermediate disturbance hypothesis are only half tested. The drop in number of species at low disturbance rates was correlated with a reduction in the number of species germinating from the seedbank. While lack of seed input in closed *Calluna* stands may partly explain the loss of species in old stands, most species also sprout after fire. Suppression of other species by *Calluna*, rather than vital attributes, may cause the loss of species at law disturbance rates.

Van Wilgen and Forsyth (1992) tested the idea that vital attributes alone would generate a quadratic relationship between richness and disturbance frequency. From an analysis of fynbos vital attributes they predicted a decline in richness with long return intervals because of the loss of early successional species with short-lived seedbanks. A smaller decline would occur at short fire frequencies because of the loss of slow-maturing, non-sprouting species. These species, mostly Proteiod shrubs, contribute little to total diversity but greatly to community biomass. Richness would be highest at intermediate frequencies of fires every 10–30 years.

In grasslands, patterns of diversity may be explained more by competitive exclusion than by vital attributes. Here most species survive fire vegetatively, and persistence of the weaker competitors may depend on regular defoliation of the taller herbs. Walker and Peet (1983) found that species richness of annually burned mesic pine–wiregrass savannas of the south-eastern United States was influenced both by fire frequency and

Table 7.8 Effects of fire frequency and a moisture gradient on species richness and evenness in a pine–wiregrass savanna (Walker and Peet, 1983). Fires were burnt annually (high frequency) or at intervals of 4 or more years (low frequency). Diversity indices were calculated from the frequency of species in ten 0.5 × 0.5 m quadrats per plot (SD in brackets). Number of plots in high and low fire treatments was two each in wet sites, three each in dry sites and eight and three, respectively, in mesic sites

Moisture class	*Species/0.25 m²*		*Evenness*	
	Frequent burns	*Rare burns*	*Frequent burns*	*Rare burns*
Wet	19.5 (0.5)	20.8 (1.7)	28.0 (2.1)	30.0 (0.9)
Mesic	22.2 (3.6)	14.0 (3.3)	31.0 (0.3)	21.2 (4.6)
Dry	13.2 (4.9)	11.0 (5.8)	21.0 (8.9)	18.7 (9.5)

location on a moisture gradient. Burning frequency had no significant effect on richness at the wet and dry ends of the gradient but annual burning significantly increased richness at intermediate moisture levels (Table 7.8). Their results support Huston's (1979) hypothesis that species richness is highest at intermediate levels of disturbance and at intermediate rates of competitive exclusion (higher in wet sites, lower in dry sites). Burning also increased richness in the tall grassveld of Natal where burnt plots had greater than 50% more grass species than unburnt controls in a 30 year experiment (Le Roux, 1989). However, richness differences were larger between plots burnt at different seasons (spring less than autumn) than at different intervals (1–3 years). Contrary results have been reported from another long-term burning experiment in the tallgrass prairies of Kansas. Here, repeated burning reduced richness relative to unburnt sites, with annual burns causing the most severe reduction (Collins and Gibson, 1990). These results are complicated by the colonization of unburnt sites by late successional forbs and shrubs.

From the few examples cited here, support for the intermediate disturbance hypothesis from fire studies is less than unanimous. Part of the problem is that diversity patterns after a single burn have been confounded with patterns created by repeated burning. To separate the two, diversity should be censused at the same stage in post-fire succession. In the grassland studies, for example, one does not know how many species persisted in dormant stages in old, unburnt stands. A review of fire effects on diversity in long-term burning experiments is long overdue. Another problem, only recently identified, is the extreme scale dependence of measures of diversity and disturbance effects on community heterogeneity.

7.5.2 Community heterogeneity

Measures of species richness are strongly scale-dependent. Species richness in small plots ('point' diversity at scales of *c.* 1 m^2) is often poorly correlated with richness in larger plots within which they are nested. This is because communities vary in patchiness or heterogeneity, defined as the dissimilarity in species composition between samples (Glenn *et al.*, 1992). High heterogeneity implies greater patchiness. The intermediate disturbance hypothesis has mostly been applied at the scale of point diversity, but the effects of disturbance on diversity at larger scales are largely unknown. In the wiregrass savanna example (Walker and Peet, 1983), annual burning increased species by 58% in 0.25 m^2 quadrats but only by 11% in 625 m^2 quadrats. The reason for the discrepancy is that burning also affected intra-community heterogeneity – but in the opposite direction. Infrequently burnt mesic grasslands were more heterogeneous than annually burnt sites (Walker and Peet, 1983).

In a recent pioneering study, Collins and co-workers used an extensive data set from long-term burning experiments in Kansas tallgrass prairies to show that community heterogeneity was negatively related to fire frequency (Figure 7.11; Collins, 1992). The mean similarity of samples of sites burnt annually was greater than for sites burnt at longer intervals. The decrease in heterogeneity with frequent burning was attributed to the stimulating effect of fire on the dominant C_4 grasses. In this case, species richness at the scale of the site was significantly correlated with heterogeneity, and the effect of high fire frequency was to reduce species richness at both small and larger scales. At even larger spatial scales, there was no

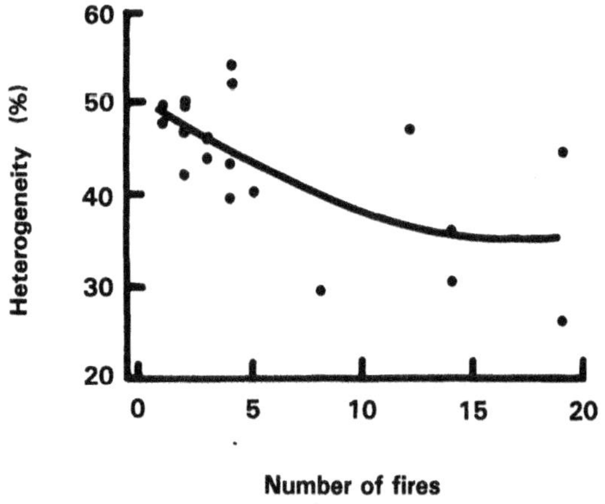

Figure 7.11 The effect of fire frequency on heterogeneity in a Kansas grassland (from Collins, 1992). Fire frequency is the number of fires burning at a site over a period of 19 years. Heterogeneity was estimated as 100 – percentage similarity between species composition recorded in twenty 10 m^2 quadrats at each site.

relationship between burning and heterogeneity, presumably because study sites began to cross community boundaries and include heterogeneity due to soil, topography and different site histories (Glenn *et al.*, 1992).

There are still too few studies of this kind to identify any general relationship between disturbance and heterogeneity or to interpret the underlying causes. Woody vegetation may show even more pronounced effects because of the heterogeneity of woody fuels and suppression of understorey species as stands age. The key question is how disturbance influences point diversity as opposed to heterogeneity within a community. We know a good deal about point diversity, very little about heterogeneity and its causes. Both scales must be considered for any complete understanding of determinants of local species richness.

7.6 WHICH RULES FOR WHICH COMMUNITIES?

7.6.1 Assembly rules

Community assembly and response rules differ among vegetation types and among growth forms within vegetation types. Vital attributes explain successional changes in fire-prone woody communities, competitive hierarchies are more appropriate where fires are rare. Grass/shrub interactions in savannas have yet another set of rules. Generalization is hazardous given the complexity of plant community ecology. It would be delightful if ecologists could locate their particular study area within a framework that predicts the key assembly and response rules. Is there any hope for a 'periodic table' for ecology which organizes information as successfully as it does for chemists? The rules developed for systems that seldom burn do not seem particularly informative for fire-prone systems. For example, Grime's (1979) well-known system of classifying plants on gradients of competition, stress and disturbance can be viewed as a set of assembly rules. The system can be used either to predict syndromes of traits expected in plants sharing the same habitat or to predict the likely habitat in which such a syndrome is most likely to occur. Grime's C-S-R system is based on vegetative attributes which, as he has observed, correlate poorly with regenerative attributes. This poor correlation has also been noted in fire-prone woodlands in Australia (Leishman and Westoby, 1992) and wetland communities of Canada (Shipley *et al.*, 1989). However, it is precisely the interplay between fire, regeneration and properties of persistence, that drives fire succession. Seed biology and sprouting behaviour are key predictors for the kinds of species likely to occur under given fire regimes. In contrast, vegetative attributes such as relative growth rate may merely define where a species can best persist within a fire successional cycle.

An alternative set of assembly rules for fire-prone communities could be derived by asking how many alternate life histories are possible under a

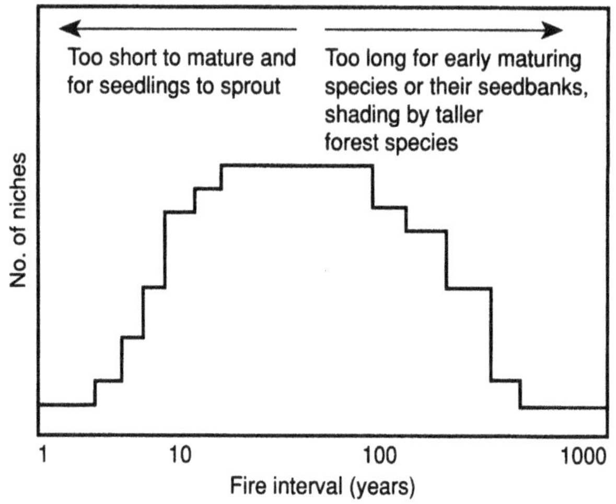

Figure 7.12 A possible relationship between fire frequency and the number of fire-successional niches. Note that the number of niches may bear no relationship to the number of species within each niche for a given fire frequency.

given fire frequency. A speculative model is shown in Figure 7.12. If fires are very frequent, few life-history alternatives are possible and sprouters will dominate. With longer intervals, there is more opportunity for partitioning the fire cycle into successional niches. For example, fire-recruiting annuals may dominate the first year after fire, followed by short-lived perennial legumes, longer-lived shrubs and then slow-maturing trees. More diversity would be feasible in the type of seedbanks, fire-stimulated reproductive modes or sprouting and non-sprouting behaviour. Where fires are very rare, as in forests or deserts, the diversity of fire successional niches would decline and be replaced by species that recruit without fire with different requirements for colonization and persistence. The number of possible successional niches within the fire cycle should therefore peak at intermediate frequencies (probably of the order of 10 to 100 years). Where fire influences only part of the community, such as the understorey of conifer forests, fewer niches would become available because the canopy remains intact and fewer resources are released by burning.

Even if this pattern of niche differentiation within the successional cycle exists, it does not tell us how many species to expect in a community. For that we would need to understand what limits the number of species occupying the same successional niche (Figure 7.13). Why, for example, are there only one or two coexisting species of ericaceous shrub in communities of European heathlands and 10–15 in South African fynbos? Why are there so many fire-recruiting legumes in Australian shrublands and so few in Californian chaparral? This is a much harder question to answer and has hardly been explored in fire ecology.

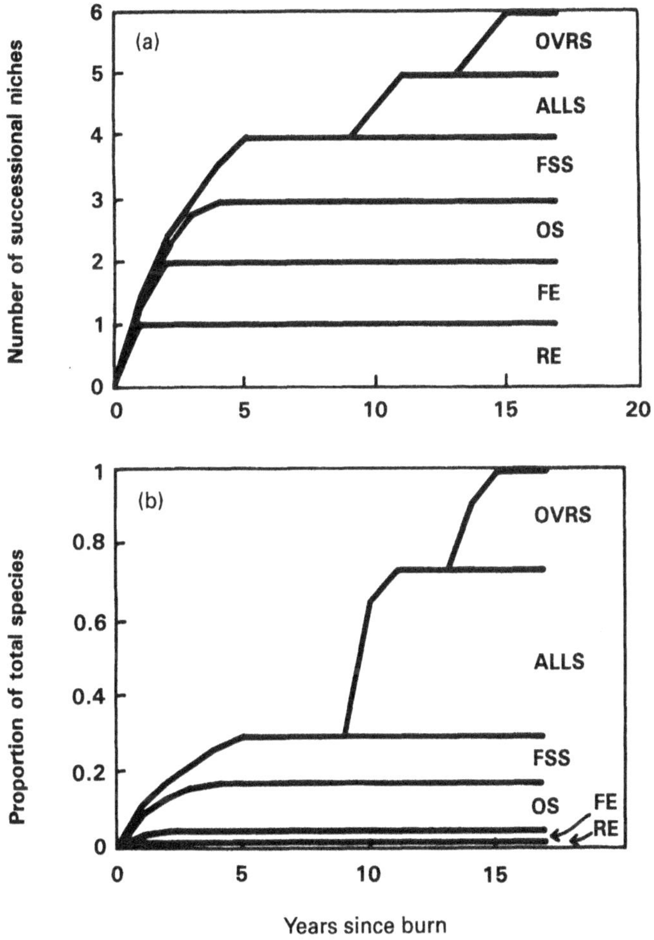

Figure 7.13 Fire interval and the number of successional niches in Western Australian kwongan (from Bell *et al.*, 1984). (a) Cumulative number of successional niches with increasing fire interval based on age to first reproduction; (b) proportion of the total species list within each niche indicating a large difference in species packing. Successional 'niches' were defined as: RE, non-fire-recruiting annuals (Chapter 4); FE, fire ephemerals (fire recruiters with life spans from less than 1 to 5 years); OS, 'obligate seeders' (fire-recruiting non-sprouters, life span greater than 10 years); FSS, 'facultative seeder-sprouter' (fire-recruiting, weak sprouters with many seedlings); ALLS, 'auto-regenerating long-lived sprouter' (fire-recruiting, strong sprouters, few seedlings); OVRS, 'obligate vegetatively reproducing sprouter' (very strong sprouters, seedlings very rare).

7.6.2 Response rules

Which response rules should we choose for which ecosystems? When is competition important in driving successional change? When is fire the more important factor? When do both need to be considered along with other key processes such as herbivory? The most ambitious attempt to answer these question stems from a paper by Hairston, Smith and Slobodkin (1960). They came to the conclusion that the answer is none of these, and that carnivores indirectly determine the structure of plant communities. They argued that carnivores compete strongly with each other in exploiting, and regulating, their (vertebrate) herbivore prey and that these therefore are not at sufficient densities to control plant biomass. Because herbivore densities are not food limited, competition should be intense among plants. Among other things, this hypothesis tries to explain why, in general, the world is 'green'. It predicts that by removing carnivores the world should turn 'brown' (or at least a different shade of green) as herbivore densities would increase and more plants would be eaten.

These ideas were expanded by Oksanen (Oksanen *et al.*, 1981; Oksanen, 1988) who argued that the intensity of herbivory would vary along productivity gradients. In habitats with low primary productivity (hot or cold deserts) there would be insufficient food to support many herbivores or carnivores. More productive habitats ('somewhat green', including grassland, mesquite, sagebrush, heath) would support a greater density of herbivores but still too few to support carnivores. Only in the most productive habitats ('truly green', forest, meadow, tallgrass prairie) would plant productivity be sufficient to maintain both large numbers of herbivores and the predator densities to control them. According to this hypothesis, the removal of carnivores would have cascading effects on the whole system in highly productive environments. Herbivore exclusion would have the greatest effect on plant communities at intermediate productivity levels. Competitive interactions would shape plant communities only at low and high ends of the productivity gradient.

Neither Oksanen, nor Hairston *et al.*, nor the many papers on the topic (see Fretwell, 1987; Oksanen, 1988; Abrams, 1993 and papers cited therein) considered fire as an additional consumer of primary productivity. The importance of fire as a consumer depends on the productivity gradient (a function of climate) but also on nutrient availability. Fires are least likely to occur where productivity is either very low (fuels too sparse) or too high (fuels too moist; Chapter 2). However, at intermediate productivity levels, Oksanen predicts that herbivores control plant communities in a so-called 'two-link' chain. Herbivore exclusion does indeed lead to rapid vegetation change in formations such as grasslands and savannas (McNaughton, 1984). So does fire exclusion. Fire and herbivory appear to be alternate 'two-link' chains whose relative importance changes depending on soil nutrient status. High levels of herbivory on nutrient-rich soils remove the fuel for burning. In such systems, exclusion of fire would have less effect on biomass than exclusion of herbivores. Low levels of herbivory, on nutrient-poor soils, allow litter – fuel – to accumulate. Here exclusion of herbivores should have less impact on biomass changes than exclusion of fire. Following the same reasoning as Oksanen (1988), competition between plants should be relatively weak in such communities. This last prediction may help explain the comparative indifference of fire ecologists to competition as an important agent structuring communities, especially at intermediate fire frequencies. We urge trophic ecologists to include fire in their models and in interpretations of the relative frequency of competition within trophic levels.

7.7 CONCLUSIONS

A central goal of plant ecology is to recognize patterns in the distribution of plants and to understand the processes and mechanisms that cause the patterns we see. Of the many factors that influence both spatial pattern and succession, interspecific competition has attracted by far the most theoretical and experimental interest. Concepts of succession, climax and the underlying processes of facilitation and competition are closely linked to ideas of stability and equilibrium – and implicit beliefs in the 'balance of nature' (Pickett *et al.*, 1992).

In fire-prone systems, nature in balance never made much sense and has long been replaced by the 'flux of nature'. Competition has been neglected as a structuring force in these systems, with the conspicuous exception of Gulliver effects in savannas. But even in grasslands and savannas, interactions within grass or woody guilds have been largely ignored. Fire ecologists have concentrated on the interplay between fire regimes and vital attributes or differences in flammability to explain patterns in space or time. They have done so at scales from patches within communities to whole continents. Their findings are backed up by some of the longest-running field experiments in ecology and are widely applied for managing natural systems (Chapter 8).

However, there is still much that is unexplained. Fire ecology is an applied science. Questions of determinants of species richness and niche differentiation have perhaps been seen as too theoretical to warrant attention. The importance of interactions in sorting assemblages of similar species has been largely ignored. There is still no general model of savannas that explains persistence of this intrinsically unstable mix of grass and trees over large areas of the Earth. There are very few studies of the importance of fire in mediating coexistence between similar competitors. There is no overview of the relationship between fire frequency and species richness. Yet fire-dependent vegetation offers much scope for community-level studies. It is particularly well suited for exploring the importance of scale in understanding ecological patterns. Perhaps the greatest challenge is to attempt to integrate knowledge of processes operating at these different spatial scales as an acid test of the effectiveness of reductionism in science.

8 Fire and management

8.1 INTRODUCTION

Fire is widely used in the management of ecosystems world-wide, mainly because it is a relatively inexpensive option by which vegetation can be manipulated. For the price of a box of matches, literally, it is possible to subject thousands of hectares of vegetation to a powerful modifying force. Unfortunately, starting fires is much simpler than achieving a desired biological effect and, if incorrectly used, fire can have many adverse consequences. Exactly how, when, and where fire is applied depends on the management objectives in the area concerned and the constraints that apply to the case. There are many, and diverse, management objectives which can influence the type of fire applied in ecosystems.

- Livestock production is dependent on the quality and quantity of fodder plants, which can obviously be influenced by manipulating the fire regime to favour such plants, or to reduce the abundance of unpalatable species.
- Forestry operations use fire in a number of ways. Site preparation, fuel reduction, and manipulation of the composition of forest stands through inducing selective mortality in fire-sensitive species are all examples.
- Conservation aims are often met by applying fire, particularly in ecosystems where fire-dependent species and communities are involved. The manipulation of habitat structure to favour certain plant species, or to improve habitat for animals, are also examples.
- Fire is used in controlling invasion by undesirable plants.
- Fire is often applied to influence the hydrological cycle, for example to enhance water yield from catchments, or to prevent erosion (e.g. Scott and van Wyk, 1992). Such applications depend on the manipulation of vegetation structure in catchments, often reducing catchment cover and biomass to low levels (for example, by converting chaparral catchments to grassland; Hibbert, 1979; Bosch and Hewlett, 1982).
- Reduction of fire hazard is a common aim of prescribed burning. Here, fires would be applied frequently, and under mild weather conditions, to reduced the build-up of fuel loads and prevent intense fires under adverse conditions.

Many of the activities that would resort under the heading 'Fire management' relate not only to igniting fires but also to preventing or suppressing fires for particular objectives. Fire prevention involves strategies to reduce the risk of unplanned ignitions, as well as strategies for the early detection of fire and for communications. Fire suppression covers the gamut of activities using manual labour, ground-based mechanical and aerial methods, and support services. There is a large literature on technical and ecological aspects of fire management. It is not our intention to review this literature here. Rather, we consider briefly the options regarding the use of fire that are open to applied ecologists who wish to use prescribed fire to manage ecosystems.

8.2 FIRE AS A VERSATILE MANAGEMENT PRACTICE

8.2.1 Constraints to burning

Fire is not a binary event (Chapter 5), implying that more must be done than simply to decide whether to burn an area or not. But how much flexibility is there when considering the options of fire? In reality, there is plenty. Fires can be implemented at different frequencies, different intensities and in different seasons. The size of the area burnt can be varied, as can the position of burnt and unburnt areas relative to each other. Each of these options will affect ecosystems in different ways, as we have expanded upon in preceding chapters. However, there are many important constraints which apply to the different options, only some of which relate to ecological concerns (Table 8.1). These are discussed in the sections below.

8.2.2 Ecological factors

The range of biological responses to fire places constraints on its application. Any fire regime that will eliminate a desirable species will eliminate the option of applying that regime. For example, fires in winter in South African fynbos are easily controlled, but they cannot be applied because of their detrimental effects on certain species (van Wilgen *et al.*, 1990b).

When fire is used for conservation purposes, variability in the fire regime is often of importance. When selecting a fire regime (a combination of frequency, season and intensity) for conservation, there may be no 'optimum' regime for maximizing diversity. The current thinking is that a fixed regime will inevitably lead to the impoverishment of diversity by repeatedly favouring the same species at the expense of others. Several authors have argued for the importance of variation in the fire regime for ensuring species coexistence (Cowling, 1987; Hobbs and Atkins, 1988; Howe, 1994). For example, serotinous Proteaceae in fynbos show maximum recruitment after summer fires that occur at intervals of roughly 15 years (Bond, 1980, 1984; Van Wilgen and Viviers, 1985). Fixed adherence to such a regime would maximize the density of the overstorey Proteaceous shrubs, but this

Table 8.1 Constraints to the use of fire as a management practice in vegetation

Constraint	*Nature of constraint*	*Areas where constraints apply*
Ecological	Fires may have undesirable effects on vegetation composition if applied at the wrong frequency, intensity or season	Areas sensitive to fire regimes, or where introduced or other undesirable species that are favoured by fire occur
Safety	The danger that prescribed burns could escape to areas which should not burn	Areas that abut on urban development, agricultural land or plantation forestry
Temporal	Seasonal or annual differences in weather result in conditions that are either too dangerous (fires become uncontrollable) or too wet (fires cannot spread)	Most areas
Legal	The application of fire may be prohibited or subject to severe restrictions	Mainly developed areas and industrialized countries
Global concerns	Increases in CO_2 and other gases emitted by fire that lead to changes in the Earth's atmosphere	All areas

could occur at the expense of understorey species (Cowling and Gxaba, 1990). On the other hand, short intervals between fires would eliminate serotinous Proteaceae. Theoretically, therefore, vegetation structure, and biodiversity, can be manipulated by varying the fire regime over time.

Different fire regimes may be needed for different ecological objectives. For example, dormant-season burns favour tall C_4 grasses in American prairies, the preferred species for large-mammal grazing. However, the rich community of associated forbs and other plants has declined under this fire regime. Where conservation is the objective, varied regimes of burn season, fire interval and grazing patterns would be more appropriate (Howe, 1994).

8.2.3 Safety

Safety is probably the most important constraint in prescribed burning. Fires, once started, can become unmanageable and cause damage to property or even loss of life. Earlier fire managers used the term 'controlled burning' for their operations, until it became clear that few fires could really be 'controlled'. The more appropriate term is 'prescribed burning', defined as the application of fire to wildland fuels under specified conditions of weather, fuels and topography so that specific objectives are achieved safely (Green, 1981).

8.2.4 Temporal constraints

Temporal constraints to prescribed burning are of two sorts. First, conditions may be such that fires simply cannot burn at certain times because of wet, cold conditions, or because the vegetation condition does not permit fires. In grasslands, for example, the sward needs to die back after the growing season, thus producing a continuous fuel bed with low moisture content that can support a fire, a process known as curing. Prior to curing, the moisture content of grass swards is too high to support spreading fires, unless of course one or more fire seasons have passed without fire or significant grazing pressure, in which case the sward would have accumulated enough dead, dry material to support a fire.

Secondly, hot, dry conditions or extended drought years will constrain the application of prescribed fires for reasons of safety. Unfortunately, fire-adapted ecosystems are normally attuned to fires at these times, and deviation from burning in this period can lead to changes in the species composition of vegetation (Chapters 5 and 7). Depending on the aims of management, this could represent a serious constraint.

8.2.5 Legal constraints

Legal constraints to prescribed burning often arise from safety concerns. They vary considerably in different parts of the world, but are usually aimed at controlling damage to property or crops, pollution, or ecological effects. With regard to direct damage to property or crops, the reasons are clear. Secondly, in many areas, particularly close to heavily settled areas, smoke pollution is a big consideration and can constrain fire management. Public attitudes to burning, especially in heavily urbanized countries, are often extremely antagonistic and fed by media reports of 'disastrous' fires 'destroying' thousands of hectares of wildland vegetation. Subsequent post-burn recovery is less newsworthy. Public hostility to burning further complicates the use of prescribed burning. Finally, legislation may be enacted to ensure certain ecological effects, and can vary from outright prohibition of fires, to seasonal or weather-related restrictions.

8.2.6 Global concerns about burning

There is growing concern about the role of fire (or biomass burning) in contributing to global atmospheric changes. Biomass burning includes the burning of forests and savanna grasslands for land clearing and conversion, burning agricultural stubble and waste after harvesting, and burning biomass fuel (Levine, 1991a). On a global scale, the total biomass consumed by annual burning is about 8680 teragrams (Tg) (1 teragram = 1 million tonnes = 10^{12} g) of dry material (dm). The estimated total biomass consumed by the burning of savanna grasslands (3690 Tg dm yr^{-1}) exceeds the other components of biomass burning – the burning of agricultural waste (2020 Tg dm yr^{-1}), the burning of forests (1540 Tg dm yr^{-1}) and the burning of fuel wood (1430 Tg dm yr^{-1}) (Levine, 1991a).

Biomass burning is not a phenomenon restricted to the tropics. Large fires are a common feature of the world's boreal forests. For example, the Chinese–former Soviet Union fire of 1987 destroyed more than 4.9 million hectares of boreal forests in only 21 days. The bulk of biomass burning is due to human activity and appears to be increasing with time. It is estimated that biomass burning may produce 40% of the world's annual production of CO_2 (Levine, 1991a).

Global accounting of the contribution of fires to atmospheric carbon indicates both the enormous extent of fires and, indirectly, their ecological importance. Some atmospheric chemists have proposed major changes in land use to reduce the effects of burning on atmospheric carbon with little regard for ecological consequences (Levine, 1991b). A recent plan for the reduction of global biomass burning (Andrasko *et al.*, 1991) included two major strategies. First, the frequency, area and amount of biomass burned in natural and intentional fires in various ecosystems should be reduced. This could be done by minimizing conversion of forest and savanna to other land uses, especially agriculture and pasture, through use of fire, and substituting sustainable land management practices and systems with minimal use of fire for fire-dependent or fire-proclivity systems.

Secondly, the fraction of primary energy supplied by biomass derived from natural systems, and the efficiency of its utilization, should be increased. This could be done by replacing unsustainable harvested biomass with sustainable managed (plantation) biomass as a fuel stock, and increasing energy extraction efficiencies for biomass conversion.

Three major classes of land uses are targets for reducing the frequency, area and amount of biomass burned: crop lands, grasslands and forests. Andrasko *et al.* (1991) suggested that this set of options could be tailored to a specific land use and unit (for example rural development project, conservation or production forest, or political unit), and set of biophysical and socio-economic factors. If taken seriously, these suggestions will be an additional constraint on the management of fire-prone ecosystems. They could have major effects on terrestrial plant cover at the expense, mostly, of the poorer, less industrialized countries of the world.

8.3 PREDICTING THE ECOLOGICAL EFFECTS OF FIRE

Applied ecologists are interested in predicting the outcome of prescribed fires on the structure and composition of ecosystems, so as to achieve a particular objective. This is what motivates research into fire ecology in many parts of the world. There are a number of recurring themes for which ecological information is needed in relation to the variables that can be manipulated in managed fire regimes. These include:

- Minimum or maximum periods between fires that will promote or reduce particular species (Chapters 4 and 7; the information included in the 'vital attributes' system of Noble and Slatyer (1980) is often sufficient).
- Fires in certain seasons that will have known effects can be selected or avoided (for example, fires should be timed to coincide with periods of dormancy; Chapter 5).
- The intensity of fires can be manipulated by selecting relevant weather conditions to achieve desired results (for example, low-intensity fires in stands of timber trees can reduce fuel loads without killing the overstorey trees; Chapter 5).
- Varying the size and location of a fire can influence post-fire recovery, landscape patchiness and animal range quality (Chapters 5 and 7).

Long-term experiments form the basis of practical fire management, especially in rangelands with high fire frequencies. Where fires occur at longer intervals, perhaps only once or twice in the working life of a farmer or forester, modelling becomes an important aid to decision-making. Under these conditions, the modeller can explore the long term implications of a particular management strategy or the consequences for vegetation patterns over a landscape.

For example, Keane *et al.* (1990) developed a simulation model of fire succession in North American Douglas fir forests. They were able to show that frequent, low-intensity fires (intervals of 10 and 20 years) prevented Douglas fir saplings from surviving and becoming part of the overstorey. Regular, more severe fires resulted in a decrease of western larch after 150 years, with a corresponding increase in ponderosa pine. Douglas fir slowly increased in basal area and became established in the overstorey after 200 simulation years. Only when no fires occurred did Douglas fir achieve dominance in the understorey, and eventually the overstorey. Models such as these are useful for exploring the long-term impact on ecosystem composition of alternative burning options.

8.3.1 Coping with variability

Stochastic variation in variables that influence post-fire regeneration (the event-dependent effects of Chapter 5) have proved a major problem in the application of models, both qualitative and quantitative, for resource management. These problems become particularly acute in areas of variable rainfall. In the past, deterministic successional models have underlain most scientifically based philosophies for biological resource management. In grazing systems, for example, succession was assumed to progress steadily along a continuum from heavily grazed, early successional states to a single persistent climax state in the absence of grazing (Ellison, 1960; Stoddart *et al.*, 1975). Addition of livestock was thought to bring about a steady deterioration in range condition. Rainfall acted along the same axis, with good rains driving succession towards the climax and droughts causing deterioration. The object of good management was to choose a stocking rate of grazers which balanced the opposing forces of grazing and successional tendency at the most productive point along the continuum.

There are many exceptions to this model, and drastic reduction in stocking rates has proved disappointing as a tool for rehabilitating degraded pastures (Westoby *et al.*, 1989). An alternative 'state-transition' model that explicitly incorporates environmental stochasticity was proposed by Westoby *et al.* (1989). They argued that vegetation occurs in discontinuous alternative multiple stable states. Transitions between states may be rapid and irreversible. Invasion of savanna by woody plants provides an example where the two states of grassland and closed woodland are both stable and the transition between them may be rare, sudden and difficult to reverse (Chapter 7).

The state-transition model leads to new approaches to management and research. A high priority would be placed on formulating and testing hypotheses about the various transitions. For example, particular sequences of rainfall, grazing and fire might be invoked to explain tree invasions. Managers would exploit opportunities for transitions from undesirable to desirable states (which might occur at rare intervals) and protect themselves against the hazards of transitions to deleterious states.

State-transition models allow explicit consideration of stochastic effects, such as drought or flood years, as well as contingent effects which might influence vegetation transitions after an event such as fire. Recently Starfield *et al.* (1993) have developed 'frame-based' models explicitly for simulating communities with state-transition attributes. The key components of a frame-based model are:

- The temporal system dynamics is partitioned into a set of states or frames.
- Independent models are constructed for each frame. These models simulate the processes that have been identified as the key processes of interest within that frame.
- Rules are established for switching from one frame to another – the transition rules (Starfield *et al.*, 1993).

An example is illustrated in Box 8.1. These models seem to offer much promise for fire-prone systems. They allow great flexibility for model construction. Deterministic models of internally driven dynamics, such as those explored in Chapter 4, can be used to describe system behaviour within a frame, while qualitative or rule-based models may be more appropriate for event-dependent transitions (Chapter 5) between frames. This flexibility of model construction allows effective use of the cumulative experience of field biologists, and the simplicity of model structure is appealing to managers.

Spatial heterogeneity in the intensity and effects of fire is a major factor

Box 8.1 A frame-based model for savannas

Frame-based models have been developed by Starfield *et al.* (1993) to simulate the state-transition vegetation dynamics described by Westoby *et al.* (1989). This approach seems to hold considerable promise for exploring management options in systems strongly influenced by events. As an example of the approach, Starfield *et al.* developed a model of the dynamics of a savanna in Zimbabwe. The objective was to draw on available ecological experience, and limited data, to compare the effects of various management strategies on vegetation over time scales of 25–100 years. The model was also used to gain insight into interactions between those processes and events thought to be important in the dynamics of the system. Three frames were defined: a grassland frame with 'Gullivers' (stunted, multi-stemmed shrubs, Chapter 7), a woodland frame dominated by *Brachystegia boehmii* with a grassy understorey and a bushland frame dominated by woody *Combretum* species with a sparse grass cover. The driving variables were elephant densities (four classes), rainfall (high/low), fire (on/off) and burn time (early/late). Fires could be cool or hot, depending on fire season and the available fuel. Transition rules for the three frames are summarized below:

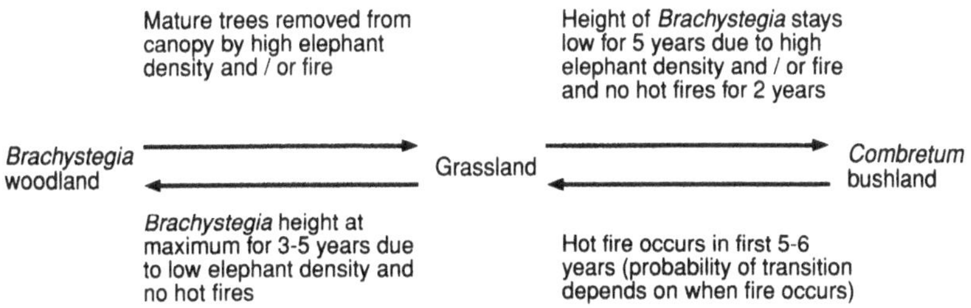

Rules were devised for dynamics within each frame and, depending on conditions in each simulated year, predicted the likelihood of transition to a new frame. For example, 'Gullivers' in the grassland frame grew taller at a rate depending on annual rainfall. Their height could be reduced by elephant browsing or fire, depending on plant size, elephant density and fire intensity. In the absence of intense fires, or heavy elephant browsing, 'Gullivers' eventually reached a threshold height from which their growth to tree height was irreversible. At this point the frame shifted from a grassland to a *Brachystegia* woodland with a new set of rules. Frame-based models can be developed with quantitative, or mixed qualitative and quantitative, rules reflecting knowledge of the system. In this relatively simple example, the

(continued)

frame-based model proved particularly valuable in exploring interactions between driving variables. For example, it showed that the system would nearly always revert to an unfavourable bushland state at high elephant densities or under higher rainfall scenarios, regardless of the use of fire.

in determining fire effects (Chapter 7). The problem of simulating spatial variation in fire effects has only been addressed relatively recently, but progress should be rapid with developments in computer simulation (Green *et al.*, 1991). The incorporation of spatial information into management models has been facilitated by the development of geographic information systems (GIS). One example is the system currently being developed for fire management of fynbos vegetation in the mountains of the Cape (Richardson *et al.*, 1994; van Wilgen *et al.*, 1994). This system includes applications for the prioritization of areas for burning, monitoring the success of fire management, mapping fire hazard for fire-control planning and the production of management summaries and statistics. Similar GIS models for fire management have been developed in Australia (Kessel, 1990) and North America (Baker, 1992). Potentially, applications of this nature will go a long way to providing a means for dealing with variability in the landscape-scale decisions that face managers on a daily basis.

8.3.2 Using fire for managing ecosystems

Once the key elements of system response to fire are determined, the application of general rules will depend on the aims of management. Some examples are provided below.

Fuel reduction

Fuel-reduction burns must reduce fuel loads (usually dead plant material that accumulates between fires), but often there will be a requirement that desirable species should not be harmed. A knowledge of the effects of fire intensity on fuel reduction, and on specific species, will be required.

A word of warning – fuel-reduction burning in the hope of preventing vast, high-intensity wildfires does not necessarily work too well in extensive fire-prone ecosystems. The fires in Yellowstone in 1988 are a case in point (Pyne, 1989; Schullery, 1989). Low-intensity fuel-reduction burns simply did not remove enough dead material, and occasional higher intensity fires were suppressed, with the inevitable result of fuel build-up and conflagration.

Encouraging specific species

Quite often, a detailed understanding of species response to fire is restricted to well-studied species that are important either because they dominate ecosystems (e.g. chamise in chaparral), or because they are problematic (e.g. invasive woody species in South Africa; Richardson *et al.*, 1992), because they are rare or otherwise of special interest, or economically important (e.g. certain pasture grasses). In such cases, the effects of varying the fire regime on the species concerned will be predictable, but the effects of the fires on co-occurring species may not be.

Maximizing biodiversity

In cases where biodiversity should be maximized (for example in conservation areas subject to recurring fire), application of a varied fire regime over time and space can be used to create diverse conditions that will not lead to a single species becoming dominant.

Enhancing water yield

Fire is used in many catchment areas to reduce biomass and thereby increase water yield. Such management can even take the form of vegetation 'type conversions', for example replacing forest with grasslands, and maintaining these using frequent fire.

Another word of warning – the reduction of biomass through burning can also lead to erosion problems in some areas. Massive erosion after the Yellowstone fire altered the physical landscape and caused abundant deposits of sediments in alluvial fans (Meyer *et al.*, 1992). In just one sediment transport event after this fire, debris was deposited for a distance of 12 km over the valley bottom. Similar fires have triggered similar massive erosion and sedimentation in the region for thousands of years.

Improving forage production and quality

Fire is an integral part of pasture management world-wide, and is applied to remove unpalatable material from the sward. Pasture managers must be cautious of not succumbing to the temptation of applying fire out of season to induce a flush of palatable growth at the longer-term expense of depleting the palatable grasses (e.g. Stoddart *et al.*, 1975; Tainton, 1981).

Manipulating the relative composition of stands

There are many examples of cases where fire is used to manipulate the species composition in vegetation, for a variety of reasons. Some examples are discussed here.

In the pine forests of the south-eastern United States, prescribed burning is used to control hardwoods, prepare mineral seedbeds for pine seedlings, reduce wildfire hazard and reduce losses to pathogens (Wright and Bailey, 1982). In the absence of fire, succession proceeds to a hardwood forest unsuitable for conifer recruitment. Prescribed burns are carefully designed to minimize damage to the pine overstorey. Different prescriptions are used for different pine species and for their different life-history stages.

Fire is used to control dieback in the jarrah forests of Western Australia by manipulating the composition of the understorey. Dieback is caused by a soil-borne fungus, *Phytophthora cinnamomi*, which attacks the fine feeding roots of jarrah and many other native plants, causing death. Logging, followed by repeated low-intensity fires, stimulated the development of stands of *Banksia grandis*, a species highly susceptible to dieback. On the other hand, less frequent, higher-intensity fires favour the germination of leguminous shrubs, and these are less susceptible to dieback. Fires are now applied with the express aim of favouring the leguminous plants to restrict outbreaks of dieback (Underwood and Christensen, 1981).

Fire is used in the control of alien invasive trees in South African fynbos shrublands. The trees (American and European *Pinus* species, and Australian *Hakea* species) are killed by fire and proliferate by means of wind-dispersed seeds. The trees are felled and left to lie prior to burning, during which time the seeds fall to the ground and are either eaten or germinate. Prescribed burns applied about a year after felling are effective in killing the seedlings before they mature, allowing indigenous species to reproduce in the normal way. The technique is very successful, but is labour intensive and expensive (van Wilgen *et al.*, 1992).

Manipulating habitats

Fire can be used to create or maintain habitats for desired animal species either by holding succession at a particular stage, or creating a mosaic of habitat types with resultant ecotones, or encouraging particular food plants. In Scotland, for example, heathlands are burnt in strips about 30 m wide to promote red grouse (*Lagopus lagopus scoticus*) populations (Hobbs and Gimingham, 1987). The grouse feed mostly on young shoots of *Calluna* maintained by frequent burning. However, older vegetation is used for nest sites and as cover from predators. The birds are territorial so that the optimal habitat is a finely divided mosaic of young and old heathland. Where fire is used to promote grouse populations, 30 m wide strips are burnt. The burn width is determined by the maximum foraging distance (15 m) from tall shrub cover.

These examples of management aims illustrate just some of the ways in which fire is used to manage ecosystems.

8.4 MANAGING FIRES

There are many ways of applying fire in practice. These range from the traditional approaches derived from centuries of human experience still used in many parts of the world, to the highly technological approaches adopted by some industrialized countries. The former are generally rather poorly documented, while many excellent texts exist for the latter (Luke and McArthur, 1978; Wright and Bailey, 1982; Chandler *et al.*, 1983; Pyne, 1984; Biswell, 1989). The amount of effort that is put into the application and control of a prescribed fire is proportional to the constraints (Table 8.1) and to the resources available. The technological approaches adopted in the more industrialized countries reflect both the value of infrastructure and crops that could be lost in fires, and legal liabilities.

When the need for prescribed burning arises, the decision of where and when to ignite the fire presents the first obstacle. Once the fire has been initiated, the task becomes one of maintaining control. These two aspects are discussed below.

8.4.1 The decision to ignite

The decision to ignite vegetation must consider whether the resultant fire will achieve the desired effect, while still remaining within the required limits of safety. The many factors affecting the potential behaviour of a fire (Table 8.2) need to be integrated in order to predict potential behaviour, clearly not an easy task. Experience provides one way of doing this, and traditional burning practices based on experience alone still account for most burning. In industrialized societies, experience has been supplemented by models of fire behaviour.

The models that have been developed are of two kinds – empirical and physical. The McArthur models (Noble *et al.*, 1980), based on many observations of fires in the eucalyptus woodlands and grassland of Australia, are good examples of empirical models. Managers in Australia use these widely for fire danger forecasting and as a guide to fire behaviour.

A good example of a physical model is the Rothermel fire spread model (Rothermel, 1972). This model (Box 8.2) has been used to support the BEHAVE fuel and fire modelling system (Burgan and Rothermel, 1984; Andrews, 1986; Andrews and Chase, 1989) used by the US Forest Service. The system is used to characterize the fuel properties of vegetation, and to predict fire behaviour under selected combinations of fuel moisture, wind and slope. Examples of model output are given in Table 2.3. It also simulates important variables such as fuel moisture contents, and spotting distances, given selected conditions. The BEHAVE system represents one of the most advanced tools available to fire managers today.

Both systems are useful in assessing prevailing and future conditions and

Table 8.2 Important variables to be considered when deciding to ignite a prescribed burn

Variable	*Methods for assessment*	*Importance for control of fire*
Wind	Measured on site, taking into account the effects of topography, and forecast conditions for the predicted duration of the burn	One of the most, if not the most, important factors in determining the potential behaviour of a fire, and therefore its relative controllability. Increases in wind lead to exponential increases in rate of spread and intensity
Temperature	Measured on site, and should take into account forecast conditions	Temperature is important due to its influence on relative humidity (see below)
Relative humidity	Measured on site, and should take into account forecast conditions	Important in determining the moisture content of fine dead fuel in the vegetation array. Fuel moisture, in turn, affects rate of spread and intensity
Topography	Assessed from local knowledge and available maps and aerial photographs	Slope has a direct effect of the rate of spread of fires (fires burn faster upslope than downslope). Topography also has an effect on wind
Surrounding fuel conditions	Assessed from local knowledge	The risk of escaped fires is related to the amount and condition of fuels surrounding an area to be burnt
Previous rainfall	Assessed from local records	Rainfall affects the moisture contents of dead fuels. In vegetation with heavy, dead fuels, the effects of rainfall can last for days or even weeks
Predicted fire behaviour	Several models exist. These include mathematical and empirical models, and are available on hand-held calculators, slide rules and tables	The ability to predict fire behaviour accurately will enhance the ability to decide on the right time for ignition, and lessen the risk of runaway fires

Box 8.2 The BEHAVE system – an example of a fire behaviour prediction and fuel modelling system to assist ecosystem managers

The BEHAVE system is a set of interactive computer programs that enable managers to:

- describe vegetation in terms of the fuel properties needed as inputs to fire behaviour prediction models, from simple field measurements;
- refine these 'fuel models' so that they produce reasonable fire behaviour predictions;
- predict the rate of spread and intensity of fires under a range of selected topographical and weather conditions;
- estimate the area and perimeter of fires, potential spotting distances and scorch height and mortality in overstorey trees;
- model fuel moisture contents to estimate potential changes in fire behaviour over a 24 hour period.

The system runs on both PCs and hand-held calculators, and is widely used by ecosystem managers, especially in the United States. The system is based on a fire behaviour prediction model developed by Rothermel (1972). Once managers have developed fuel models for their vegetation, only estimates of fuel moisture content, wind and slope are needed as inputs.

their effects on fire behaviour, thus enabling more informed decisions to be taken when deciding whether to ignite a fire or not. The models are also extremely useful training tools, enabling managers to grasp the principles of factors that affect fire behaviour.

Despite the existence of these models, and a body of ecological knowledge, managers are often called upon to make decisions on whether or not to burn in the absence of experts who can advise them on an appropriate course of action. One solution to this problem is through the use of expert systems. Expert systems are computer programs capable of holding an apparently intelligent conversation with the user. Based on the answers to the questions, and the data and rules held in the system, they help the user to arrive at a decision and to track the logic of such a decision. The problems of availability of expertise to managers may result in the wider use of expert systems in land-management decisions in future (Noble, 1987).

Starfield and Bleloch (1983) introduced expert systems to ecology by describing such a system to advise on prescribed burning in African savannas. Because of the potential of expert systems for dealing with situations where sound advice is needed quickly (such as when dealing with fires) it is not surprising that a high proportion of the first examples in ecology dealt with fire management (Starfield and Bleloch, 1983; Walker *et al.*, 1985; Ludwig, 1990). They certainly have the potential to improve the quality of fire-management decisions. However, it is too early to attempt a review of their success in this regard. At this stage, expert systems still represent a challenging opportunity to researchers, in which they need to express their ideas in concise, logical rules (Noble, 1987), and to make such knowledge readily available to managers.

8.4.2 Maintaining control

Once a fire has been initiated, the problem becomes one of maintaining control. Maintaining control of an active fire front is a difficult task, but there are principles that can assist managers. Features of the terrain such as roads or major rivers, firebreaks or recently burnt areas can be used to advantage in containing fires. Changes in wind can play havoc with the ability to control free-burning fires, but can be used to advantage in combination with backfiring to contain fires within a desired area. Humidity, which affects fuel moisture, is another important variable to monitor.

Changes in wind and humidity, together with conditions of atmospheric instability, can lead to spotting. Spotting is the development of new fires ahead of the main fire front by firebrands carried aloft by wind and is a very dangerous condition for prescribed burning.

Recent developments in the use of geographic information systems (GIS) to support dynamic spatial modelling have resulted in an embryonic ability to simulate the growth of fires over the landscape (Green, 1989; Ball and Guertin, 1992; Vasconcelos and Guertin, 1992). Such developments hold great promise for the future, and some are already capable of real-time predictions (Green *et al.*, 1991).

Once a prescribed burn has been completed, securing the safety of the perimeter becomes a priority. Finally, flare-ups even some days after completion of burning operations, have to be guarded against.

8.5 CONCLUSIONS

Fire is an integral part of the dynamics of the Earth's surface, and the world's biota have evolved to cope with the phenomenon of recurrent fires. Fire is so essential to the maintenance and manipulation of ecosystems that it has to be considered and used by all managers of such ecosystems. Despite the concerns about the effects of fires on the dynamics of the atmosphere (mentioned above), there can really be no serious consideration of the elimination of fire from major portions of the Earth's surface to alleviate these concerns. Fire is necessary for the maintenance of biological diversity and the production of livestock. It is, in any case, an inevitable consequence of a combination of fuel, weather and sources of ignition.

Managers who have attempted to exclude fire from ecosystems in the past have learnt the folly of their actions the hard way.

Fire is used to manipulate ecosystems despite an incomplete understanding of the consequences in many cases, and it will continue to be used. The philosophy of managing for some fixed 'optimum' state is gradually being replaced by one that recognizes the benefits of diversity in the use of fire. Much progress has been made towards developing an understanding of the effects of fire, and with increased knowledge has come the realization that prediction of both the behaviour and effects of fires must rely on either a vast degree of experience or complex models. Models have proved useful as training aids, or as aids to enhancing understanding of the complexity of ecosystems subject to fire. Further developments in computer technology hold the promise that models will become increasingly useful in the application of fire and the prediction of its effects. However, there is intrinsic uncertainty in the behaviour and effects of fire, and we can anticipate many surprises.

9 Fire and the ecology of a changing world

9.1 INTRODUCTION

Most of this book has dealt with fire effects at the scale of populations and communities, with some consideration of landscape-level effects. In this chapter, we look at the large-scale impact of fire on global vegetation and the evolution of floras. We pose the hypothetical questions of what the vegetation of the world might look like in the absence of fire and how the inclusion of fire might alter systems which now seldom burn. Would there be radical changes in structure? Would grasslands be replaced by forests without burning, or forests become grasslands as fires intrude? We develop the chapter in the context of two key social concerns that drive an enormous amount of contemporary research effort: global climate change attributed to human impacts on global atmospheric chemistry and the imminent threat of extinction of much of the world's biological diversity. They provide a useful framework for discussing the contributions and the shortcomings of our current ecological understanding in general and of fire ecology in particular.

The threat of global warming is forcing ecologists to re-evaluate causes of large-scale biogeographical patterns in order to predict how the distribution of organisms will change and at what rate. A strong correlation between global climate and vegetation formations appears, at first sight, as good evidence for direct climate control on plant growth. We first explore such correlations and the evidence for direct climate control. Climate may influence vegetation indirectly through its effects on fire. However, untangling the respective effects of climate, fire and vegetation is difficult. We first discuss palaeoecological and contemporary evidence that supports a strong relationship between climate and fire. Strong fire/climate correlations may suggest that the fire regime is an inevitable consequence of particular climates. However, there are feedbacks between vegetation and fire and not only between fire and vegetation. The same climate may support both flammable and non-flammable formations. Fuel properties may have evolved so that some formations burn more readily than others. We explore the evidence for fire acting as an evolutionary force in moulding floras. If species have evolved fire-dependent traits, then vegetation may be more than a passive player in the fire/climate interaction. Different fire regimes may occur under the same climate depending on the evolutionary history of the floras.

Climate/fire interactions and their effect on vegetation distribution may also depend on the frequency of ignition events and the type and continuity of fuel. Human use of fire has altered ignition patterns over the globe. Is there evidence for major vegetation transformation due to human activity in support of large-scale impacts of fire? Human settlement on remote islands forms a 'natural experiment' for exploring such effects, and we discuss some putative examples. Human impact on fires and fuels have caused many changes to the interaction between fire and vegetation. We discuss the nature and magnitude of these changes. For example, the fragmentation of landscapes by cities, farms and roads has changed the continuity of fuels, either reducing or enhancing fire frequency depending on circumstances. Finally we ask how knowledge of climate/fire interactions and vegetation responses can be used to predict both the impact of global warming and changing disturbance regimes on vegetation change.

9.2 CLIMATE, FIRE AND BIOGEOGRAPHY

9.2.1 Climate and the distribution of vegetation

The close relationship between major vegetation formations and climate first emerged during the explorations of the great German biogeographers in the nineteenth century. What particularly impressed these early observers was the convergence of vegetation structure in areas of similar climate but different phylogenetic histories. The evergreen sclerophyll shrublands of Mediterranean-type climates are classic examples (Mooney, 1977; Cody and Mooney, 1978). Correlations between climate and vegetation formations still form the basis of many attempts to predict changing vegetation patterns from predicted changes in global climate. While these can give us a good general idea, the underlying assumption of these studies is that the correlation is also causal and that climate directly determines the distribution limits of vegetation formations. However the closer one looks for direct climate impacts on plant distribution, the harder it is to find them. Darwin observed, 'the change of climate being conspicuous, we are tempted to attribute the whole effect to its direct action.... but this is a false view. Climate acts in the main part indirectly ...'. Experimental studies have largely supported Darwin's view, and convincing tests of direct climate control of species limits have proved rare and elusive (Woodward, 1987). The implication is that species responses to climate change will depend at least as much on the set of interacting species as on absolute physiological limits. The importance of interactions in determining the realized, as opposed to the fundamental, niche are difficult to reconcile with broad-scale correlations between climate and vegetation.

To what extent are climate/vegetation correlations due to indirect effects of climate on other ecological processes? Two other globally important processes include the controlling influence of higher trophic levels (Oksanen *et al.*, 1981, Fretwell, 1987; Chapter 7) and of fire, including fires of anthropogenic origin (Sauer, 1950; Stewart, 1956; Jackson, 1968; Wells, 1969, 1970; Axelrod, 1985). Defoliation levels by both fire and herbivory are influenced by climate. However, it is important to separate direct and indirect effects of climate. For example, regional climate may be conducive to frequent grassland fires but support mosaics of grasslands and woodlands because of obstacles to fire movement such as large rivers (Grimm, 1984). In such cases, suppression of fires can initiate succession from grassland to woodland with no change in climate. How does one separate the direct effects of climate from its many indirect effects on other ecological processes?

Woodward has developed a promising new approach that partly escapes the tautology implicit in many climate/vegetation correlations. He developed a mechanistic model which predicts maximum leaf area possible under a given climate from the physical constraints of available energy and moisture on plant growth (Box 9.1). The predicted leaf area index (LAI) was then used to infer vegetation structure using correlations with present-day vegetation which indicate that the greater the LAI the taller and more forest-like the vegetation (Woodward, 1987, 1992). The power of this approach is that it can be used as a kind of 'null' model for exploring the deterministic effects of climate on vegetation. Where vegetation in a region is consistent with model predictions then plant growth can be assumed to have reached its full climatic potential. Large discrepancies, such as grasslands occurring where forests were predicted, would indicate the

Box 9.1 A physically based model for predicting leaf area

It would be useful to know how much plant growth could be supported under a given climate from a purely physical point of view. This theoretical maximum could be compared with actual plant cover to help separate direct and indirect effects of climate on vegetation. Woodward (1987, 1992) has developed such a physically based model to predict leaf area index (LAI), the total amount of leaf area per unit of ground. LAI is measured in units of leaf layers so that an LAI of 3 represents a canopy of three leaf layers. A high LAI adds to the water stress of plants in two ways, by increasing transpiration and by decreasing the amount of precipitation reaching the soil (throughfall). Transpiration rises with increasing LAI because the greater the leaf surface area, the more water can be transpired. Throughfall declines with increasing LAI because more rain is intercepted by the leaves and evaporated before reaching the soil. Throughfall and

(continued)

transpiration for a given climate can be calculated from physical principles using energy balance models. The maximum LAI possible for a given climate is determined as the point where water loss from transpiration balances the amount of water available to the plant from throughfall (see figure below). The theoretical maximum can be compared with measured LAI to determine whether the vegetation in a region has reached the physiological limits imposed by the climate. LAI correlates with vegetation height (taller plants can fit in more leaf layers), providing a crude estimate of potential vegetation structure for a given climate (Woodward, 1987, 1992).

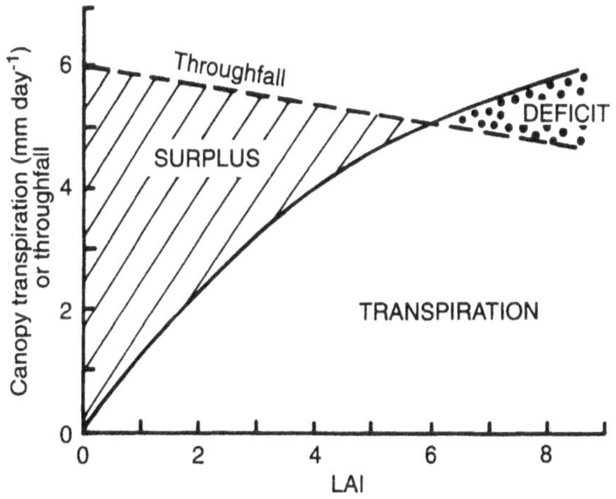

Predicted throughfall and transpiration in mm day^{-1}, for canopies differing in leaf area index (LAI). The maximum LAI that can be supported in this example is 6, given by the intersection of the two curves. At LAI below this maximum, surplus moisture would be available (hatched area). At LAI greater than 6, transpirational demand would exceed the throughfall (dotted area) (modified from Woodward, 1992).

overriding importance of other factors such as fire, herbivory or soil nutrients (Plates 5 and 6). Global vegetation patterns derived from the LAI model have been compared with global vegetation maps with some success (Woodward, 1992). Unfortunately, the resolution (just three vegetation types to describe global vegetation) is too coarse to identify interesting anomalies. Nor is it clear just how much can be inferred from LAI about vegetation structure.

Global climate/vegetation correlations are misleading in implying direct climate control of vegetation. Mechanistic approaches, such as that of Woodward, offer more rigorous solutions to the problem of separating climate from other factors, including fire, as determinants of major biome patterns.

9.2.2 Fire and the distribution of vegetation

Many ecologists have argued that the major fire-prone formations, especially grasslands and savanna, owe their character and geographic extent to regular burning, especially anthropogenic burns (Sauer, 1950; Stewart, 1956; Wells, 1970; Booysen and Tainton, 1984). The implication is that the impact of global climate change on vegetation would be mediated by fire and could be moderated by managing fire regimes. The physiognomic tendencies of regular burning are:

- to reduce vegetation height (tall forests to shorter ones, woodlands to shrublands);
- to replace woody vegetation by grasslands;
- to promote flammable species or communities (low litter decomposition rates, more xeromorphic leaves, finer twigs/branches; Chapter 2);
- to reduce biomass.

Thus fire-affected communities should be of lower stature and more open than predicted, say, from Woodward's climate/LAI models. Exclusion of fires should allow vegetation to reach its climatic potential through replacement of grassland by woody plants, shrublands by trees and woodlands by tall forests. The argument for the importance of fire has been based on the frequent occurrence of alternative flammable and fire-free vegetation types in fire-prone landscapes (Plates 5 and 6), and experimental demonstration of transitions between the two when fire is excluded (Chapter 7). For example, closed forest patches in savanna and shrubland landscapes may represent the climatic optimum, but the regional vegetation cover is a product of the fire regime. At issue, however, is not the fact that fire can drastically alter local vegetation patterns but the scale on which it does so. What proportion of British heathlands, for example, could support conifer forests or deciduous woodlands if fires were suppressed and tree seed sources still available? These questions need to be addressed when teasing out the role of fire from climate in determining the distribution limits of major vegetation types.

9.2.3 Climate/fire correlations

Is fire regime an inevitable consequence of a particular climate? A number of studies have explored the correlation between climate and fire at time scales from millennia to decades or even cycles of smaller duration such as those associated with El Niño southern oscillation effects. These show that fire regime is closely correlated with climate and that global warming will influence the frequency and severity of fires. In some fire-prone vegetation, fire/climate interactions may be of overriding importance for vegetation change.

Evidence from charcoal deposits and fire scars has been particularly useful in unravelling fire histories in the northern hemisphere. Clark (1988,

1990) was able to reconstruct fire histories of mixed conifer/hardwood forests in Minnesota using both charcoal deposits in laminated lake sediments and fire scar analysis. Using independent climate reconstructions, he was able to establish an association between fire recurrence interval and climate over a 750 year period (Figure 9.1). Fire was most frequent during the warm, dry fifteenth and sixteenth centuries (*c.* 9 year recurrence intervals) and least frequent during the Little Ice Age (*c.* 25 years in the mid-eighteenth and 44 years in the mid-nineteenth centuries). During warm

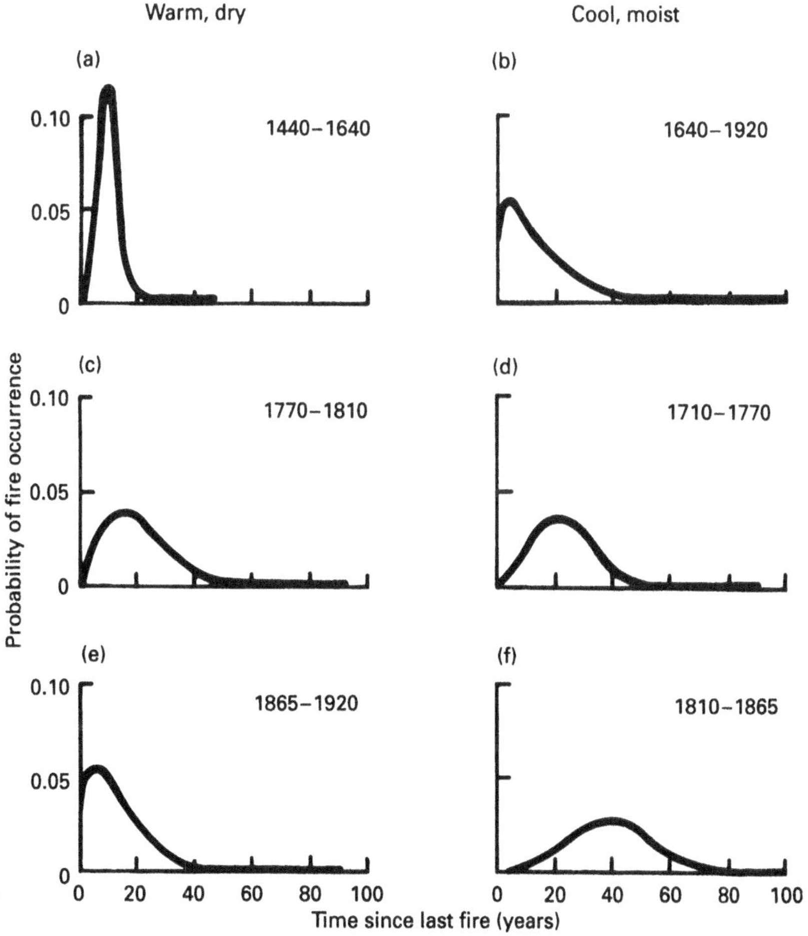

Figure 9.1 The effects of climate on the probability of fire occurrence over the past 750 years in north-western Minnesota (from Clark, 1989). (a) The warm, dry period from AD 1440–1640 was characterized by frequent fires and a low variance in return intervals. (b) With the onset of cooler, moister conditions after 1640, fire frequency shifted to a mix of short and long fire intervals. Within this period, fire frequencies were still sensitive to minor climate fluctuations. Warm, dry periods again produced higher probabilities of short return intervals (c,e), while longer return intervals occurred during cool periods (d,f).

periods, the probability of fire seemed limited only by the availability of sufficient continuous fuel in the litter layer. During cool and moist times, fire was most likely to occur in years with high moisture deficits. These and similar studies (e.g. Johnson and Larsen, 1991) are providing clear evidence for the historical association between climate and fire regimes.

Palaeoecological studies are rarer in southern-hemisphere vegetation, partly because few trees have annual rings. However, the relationship between climate, fire and vegetation has also been demonstrated in Australian eucalypt woodlands. For example, Hopkins *et al.* (1993) used charcoal deposits to trace the extent and timing of transitions between eucalypt woodlands and humid rainforests in northern Queensland. The location of charcoal fragments indicated that sclerophyll forests penetrated what are now some of the wettest rainforest areas in the period from 12 000–8000 BP. This period coincided with increased aridity in the Late Pleistocene. Reinvasion of rainforest since 8000 BP coincided with wetter and less seasonal rainfall than occurs at present. Aboriginal burning practices have also been implicated in Australia's changing vegetation, with a large increase in charcoal coinciding with Aboriginal settlement (Kershaw, 1986; Pyne, 1989). However, Aborigines were present throughout this period and reinvasion of forests occurred despite their burning practices. As in northern-hemisphere studies, the pattern of a close correlation between climate and fire regime seems to be emerging. What is more, the vegetation changes seen in the fossil record are consistent with observed effects of changes in fire frequency in contemporary vegetation (Chapter 7; Jackson, 1968).

Palaeoecological studies may be poor models for future climate/fire interactions. One problem is the huge preponderance of human-caused fires in modern landscapes. However, the largest, most intense contemporary fires are still closely associated with extreme climatic conditions. In Mediterranean sclerophyll formations, for example, the total area burned over a 15 year period was correlated with particularly hot periods (>25°C) in central and northern Spain and with drought periods along the Mediterranean coast (Vasquez and Moreno, 1993). Mean annual temperature was a poor predictor of burnt area. This relationship between fires and extreme weather conditions is also manifested as a link between fire occurrence and El Niño southern oscillation (ENSO) weather anomalies (Swetnam and Bettancourt, 1990; Brenner, 1991). The large-scale fires of 1982–83 in the whole of the western Pacific region were a consequence of regional droughts associated with ENSO phenomena. On Borneo alone, 5 million ha of rainforest vegetation were affected by fires spreading from numerous point sources throughout the island in this period (Goldammer and Seibert, 1990). Attempts to model the effects of global climate change on fire properties (e.g. Torn and Fried, 1992) have been hampered by the poor resolution of current climate models with respect to temperature extremes and rainfall seasonality, both key predictors of fire occurrence. However, the indications are that global warming will lead to more frequent, more intense fires in the future, just as happened in hot, dry periods in the past.

Both palaeoecological and contemporary studies indicate that fire regimes closely track climate. Global warming may similarly cause rapid changes in fire regimes in the future and also trigger rapid vegetation change.

9.2.4 Fire and the evolution of floras

Palaeoecologists commonly assume that all vegetation change is caused by climate change. Climate/fire correlations provide some support for this single-factor view of vegetation change. However, fire propagates only in appropriate fuels. Is fuel type also an inevitable product of particular climates, or have components of plant formations evolved to become more flammable? In other words, are plants passive responders to their physical environment or have they evolved attributes that change physical conditions for survival? Mutch's (1970) hypothesis, that traits enhancing flammability have evolved and help maintain fire-prone communities, implies that changes in the fire regime will lead to substantial changes in vegetation composition. The hypothesis has been severely criticized (Chapter 6), with the unfortunate result that the importance of fire in shaping the evolution of floras has been neglected. For example, convergent evolution in Mediterranean-type sclerophyll formations has been studied in the context of climate and nutrient constraints on plant growth but with little reference to how fire has influenced plant form.

Mediterranean-type floras

It is interesting to consider how the structure of Mediterranean-type shrublands might differ in the absence of fire. All Mediterranean regions seem to contain both species adapted to fire and species that merely tolerate burning (Naveh, 1975; Keeley, 1986; Le Maitre and Midgley, 1992). 'Adaptation' has been used loosely in the context of fire. However, fire-stimulated reproduction can be a good indication of an adaptive response to fire (Chapter 3). All the Mediterranean regions have some species with fire-stimulated reproduction (Table 9.1). For example, serotiny has evolved independently in all the Mediterranean systems but Chile. The discovery of smoke-stimulated germination in South Africa (de Lange and Boucher, 1990) and charate-stimulated germination in California (Keeley *et al.*, 1985) led to similar discoveries in other Mediterranean regions, though the extent and evolutionary significance of the phenomena are still being explored (Chapter 3).

All of these systems also contain species, typically trees or tall shrubs, whose recruitment does not depend on fire (Chapter 3; Keeley, 1986). These recruit in small canopy gaps or in the shade of unburnt vegetation

Table 9.1 The occurrence of fire life-history traits in Mediterranean-type regions. *, present; –, absent; ?, uncertain. Data compiled from Keeley (1986, 1991), Le Maitre and Midgley (1992), Specht (1988), Trabaud (1984) and references cited in Chapter 3

Trait	*Chile*	*California*	*Mediterranean Basin*	*Australia*	*South Africa*
Fire stimulated reproduction					
flowering					
grasses	*	*	*	*	*
geophytes	–	*	–	*	*
seed release					
serotiny	–	*	*	*	*
germination					
heat	*	*	*	*	*
charate	–	*	?	*	*
smoke	?	*	?	*	*
Fire survival (shrubs)					
basal sprouts	*	*	*	*	*
lignotubers	*	*	*	*	*
epicormic buds	*	*	*	*	*
non-sprouters	–	*	*	*	*

and are favoured by long intervals without fires (Chapter 4; Keeley, 1986; Manders *et al.*, 1992). The two groups tend to have different dispersal modes (Table 9.2). Fire-recruiting shrubs and trees typically have seeds dispersed by wind, gravity or ants, and form seedbanks. Interval-recruiting species often have nuts (oaks of various genera) or fleshy fruits dispersed to safe sites by vertebrates, and do not form long-lived seedbanks (Keeley, 1986). In the long absence of fire, Mediterranean-type communities are invaded by the vertebrate-dispersed group of trees and tall shrubs (Moll *et al.*, 1980; Keeley, 1986; Manders, 1990, Manders *et al.*, 1992). According to this argument, fire substantially determines the composition of Mediterranean-type vegetation.

The alternative fire reproductive strategies are associated with alternative modes of fire survival (Table 9.2). Interval-recruiting species are typically vigorous sprouters ('obligate' sprouters; Keeley, 1986). Sprouting is the norm in woody plants, even in forests that never burn, and occurs in response to a variety of injuries. Obligate sprouting appears to be a fortuitous phylogenetic accident allowing the persistence of these species in a fire-prone system. The resilience of Mediterranean shrublands to burning owes much to the presence of these 'obligate' sprouters (Naveh, 1975; Trabaud, 1984; Keeley, 1986). Fire-recruiting species are often

Table 9.2 Generalized traits associated with fire-stimulated recruitment and non-fire-stimulated recruitment in trees and shrubs of Mediterranean regions (modified from Keeley, 1986; Specht, 1988; Le Maitre and Midgley, 1992)

Species traits	*Fire-recruiting species*	*Non-fire-recruiting species*
Unusual germination requirements	Heat, charate, smoke	None
Regeneration timing	Mostly post-burn	Mostly after long fire-free intervals
Seed longevity	Several to many years	<1–2 years
Dispersability	Low, except in serotinous spp.	High (vertebrate or wind dispersed)
Fire survival mode	Sprouters, non-sprouters	Vigorous sprouters
Growth form	Shrubs or, in serotinous species, small trees	Tall shrubs and trees
Leaf size	Small, medium	Usually larger than fire-recruiting species
Flammability	Moderate to high	Low
Water relations	Drought tolerators	Drought avoiders (often deep rooted)

Representative taxa in the two groups are:
Fire-recruiting: Chile, *Colliguaya odorifera*, *Trevoa trinervis*, *Muelenbeckia hastulata*; California, *Ceanothus*, *Arctostaphylos*; Mediterranean Basin: *Cistus*, *Cytisus*, *Pinus halepensis*, *P. brutia*; Australia, many species in many families including Proteaceae, Myrtaceae, Fabaceae; South Africa, many species in many families including Proteaceae, Ericaceae, Fabaceae, Asteraceae.
Non-fire-recruiting: Chile, *Quillaja saponaria*, *Peumus boldus*, *Cryptocarya alba*, *Lithraea*; California, *Quercus*, *Rhamnus*, *Heteromeles*, *Malosma*; Mediterranean Basin, *Quercus* spp., *Phillyrea*, *Pistacia*, *Olea*; Australia, 'rainforest taxa' including Monimiaceae, Elaeocarpaceae; South Africa, *Maytenus*, *Rhus*, *Diospyros*, *Olea*, *Heeria*, *Rapanea*.

non-sprouters or 'facultative sprouters', representing the fire-adapted condition in these environments (Chapter 6). Fire-recruiting non-sprouters occur in all Mediterranean-type shrublands except Chile (Table 9.1).

Can a fire-dependent flora become a fire-promoting one? The most controversial evolutionary effect of fire is on plant form. Physiologists have interpreted convergence of plant form in Mediterranean shrublands in terms of climate or nutrient constraints (Mooney, 1977; Cody and Mooney, 1978; Specht and Moll, 1983). But several characteristics of Mediterranean shrubs also influence flammability including canopy architecture, leaf moisture content (sclerophylly), the shedding of plant parts and leaf chemistry (Chapter 6; Rundel, 1981; Bond and Midgley, 1995). Enhanced flammability is more likely to have evolved in fire-recruiting species than in species that recruit into mature vegetation (Bond and Midgley, 1995). These two groups have seldom been separately considered in ecophysiological studies of convergent evolution. There are indications that species within Mediterranean communities differ in their fuel properties (Papio and Trabaud, 1991), but there is no agreement on which traits most enhance flammability. It would be interesting to explore correlations between fire life-histories and vegetative traits as an alternative to climate and nutrient pressures for convergent evolution.

Chile is an interesting exception to the general pattern of fire-adapted and non-adapted guilds in Mediterranean-type shrublands. Although fires occur in Chilean matorral, it has been argued that natural fires are too rare to have led to the evolution of a fire-prone flora (Mooney, 1977; Munoz and Fuentes, 1989). However, anthropogenic fires are common in matorral and some species have heat-stimulated germination (Munoz and Fuentes, 1989). Munoz and Fuentes have argued that heat-stimulated seeds would not survive the extremely high soil temperatures they recorded in matorral burns (temperatures greater than 100°C sustained for more than 5 hours after the fire at depths of 5 cm). Certainly the Chilean matorral provides an interesting test case which may help in separating the respective role of fire and climate in the structural convergence of Mediterranean-type ecosystems.

The patterns seen in different Mediterranean regions suggest that the correlation between Mediterranean climates and evergreen shrublands is indirect. In the absence of fire, Mediterranean climates seem able to support broad-leaved forests and woodlands over much larger areas than at present. Instead, a group of species has evolved that produces flammable shrublands of lower stature. The evolution of a fire-dependent flora may also have been important in generating the rich diversity of Mediterranean systems, since the fire-recruiting shrubby genera are often richest in species (Wells, 1969; Le Maitre and Midgley, 1992; Chapter 6).

Savanna floras

The evolution of fire-prone grasses must have had an extraordinary impact on global vegetation patterns. Grass fuels create a distinctive fire regime that must have acted as a powerful selective filter, permitting access to only a small subset of woody plants from the ancestral forests. To test this kind of effect, one could predict that where savannas are of recent origin, woody taxa should be shared between forests and grassy formations. Where fire-derived systems are of greater antiquity, separate floras should have evolved with distinct savanna and forest elements. In the West Madagascar savannas, supposedly of recent origin, 28 endemic trees occur, of which only four are restricted to grasslands, the rest being 'transgressors' from adjacent forests (White, 1983). In the Zambezian region of Africa, with a similarly sized flora (*c.* 8500 species), savanna and forest floras are distinct, with large woody genera, such as *Brachystegia*, *Combretum* and *Uapaca*, and monotypic genera, such as *Colophospermum,* more or less confined to fire-prone formations (White, 1983). A rich savanna flora, distinct from forest formations, has also been recognized in South America (Sarmiento, 1983). The presence of endemic and near-endemic savanna tree genera attest to the antiquity of these formations. In Australia, the adaptive radiation of *Eucalyptus* has been similarly linked to the spread of fire-prone vegetation.

An evolutionary perspective is still embryonic in the fire literature. However, it does seem as though distinct fire floras have evolved. The fire/climate relationship is not determined by climate alone. The fires that occur within a given climate depend on the evolution or migration of flammable floras. The persistence of flammable and non-flammable formations under the same climate show that alternative solutions have evolved for the same physical circumstances.

9.2.5 Human impacts on fire and large-scale vegetation changes

Humans have used fire for over a million years (Brain and Sillen, 1988). One way of separating direct and indirect effects of climate and fire on vegetation would be to explore the effects of human settlement on pristine vegetation. However, though large-scale human impacts on fire and vegetation have been claimed in many parts of the world, the evidence is circumstantial and controversial. For example, it has been argued that Aboriginal fire practices transformed Australia's vegetation, producing the current dominance of eucalypt woodlands at the expense of fire-sensitive formations (Kershaw, 1986; Pyne, 1989). The problem with this and other 'natural experiments' is that climate has also changed over the long period of human settlement in Australia, and it is difficult to separate their respective effects.

Several oceanic islands have been settled much more recently, and here vegetation change is less likely to be confounded by climate change. Madagascar, the world's fourth largest island, was first settled by humans only some 2000 years ago. Humbert (1927) and others believed that at this time the island was covered by forests, thickets and sclerophyll shrublands (Figure 9.2a; Koechlin *et al.*, 1974; White, 1983). Through a lethal combination of shifting agriculture and burning, vast areas were thought to have been transformed into grasslands and wooded grasslands (Figure 9.2b). The recent origin of the grasslands was inferred from their floristic poverty, a surprising anomaly given the diversity of climate and soil types in which they occur. Forests surviving in fire refugia have been interpreted as remnants of the pre-settlement condition. The extinction of the Malagasy megafauna, including giant lemurs, elephant birds and the pygmy hippopotamus, was also attributed to this massive habitat modification.

There can be no question of the power of humans to modify the island's vegetation. Recent remote sensing studies indicate that Madagascar's eastern rainforests have shrunk by 50% since 1950 alone (Green and Sussman,

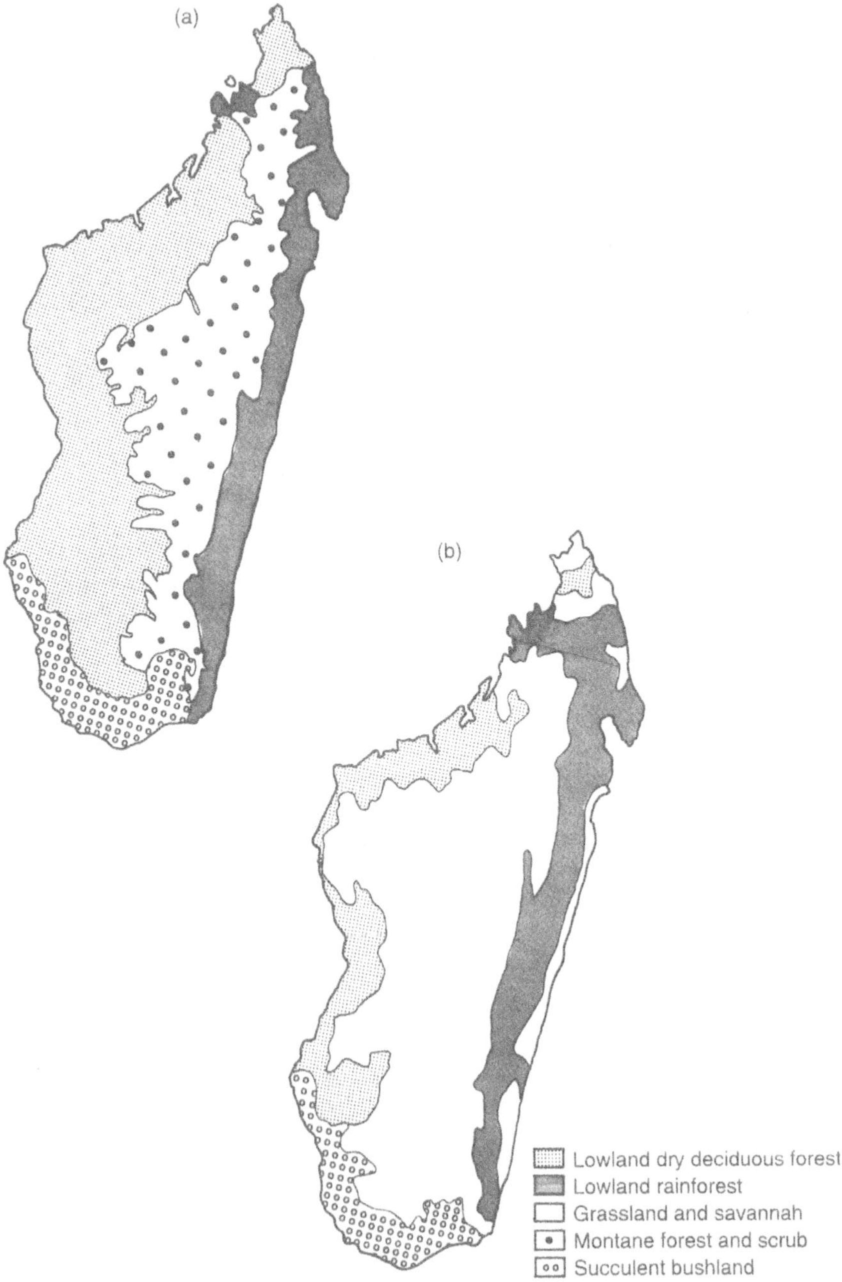

Figure 9.2 The putative spread of grassy vegetation in Madagascar following human settlement and the widespread use of fire. Recent palaeoecological studies cast doubt on the extent of human-induced changes. (a) Pre-settlement vegetation; (b) contemporary vegetation (derived from maps of White, 1983; Preston-Mafham, 1991).

1990). However, the importance of humans and fire in transforming Madagascar's great central plateau from forests to grassland is more controversial. Pollen and charcoal studies from a lake in central Madagascar indicate that open grassy vegetation and fire existed in this area for thousands of years before human settlement (Burney, 1987). Without better spatial resolution of palaeoecological data, it is difficult to attribute Madagascar's current vegetation to human use of fire. Much the same problem occurs in other large-scale examples of supposed fire-driven vegetation change.

9.3 CHANGING PATTERNS OF FIRE IN MODERN LANDSCAPES

Although the importance of fire as a determinant of global vegetation patterns is far from clear, there is no question of its significance in determining local and regional vegetation patterns. How have fire regimes changed in the past, how will they change in the future and with what ecological consequences? The pattern of fires depends on complex interactions between ignition sources, climatic conditions, type and amount of fuel and the continuity of fuels in the landscape (Chapter 2). All of these, including climate, have been greatly modified by human activity in the twentieth century. Active management of fires and variation in fire-fighting capability add to the complex intrusion of humans into the natural history of fire. Through their effects on fire, humans have had, and will continue to have, major effects on the biota.

9.3.1 Patterns of ignition

In most parts of the world, humans have replaced natural ignition sources as the main cause of fire (Table 9.3). Human influence is so ubiquitous that sources of ignition may no longer limit the area burnt except in remote wilderness areas, such as the boreal forests, or fragmented habitats surrounded by landscapes that do not readily propagate fire. However, the timing and location of human-caused fires can have very substantial ecological effects (Chapters 4, 5 and 7).

9.3.2 Changes in fuels

Palaeoecological studies suggest an inevitability about climate/fire relationships that may no longer apply in our heavily modified landscapes. For example, fuel properties have been heavily modified by human activities. Fire suppression practices over many decades in the industrialized countries have produced large accumulations of fuel and increased the risk of catastrophic fire. The scale of the suppression activities is enormous. The boreal forests of the Russian Federation cover *c.* 730 million ha, an area

Table 9.3 Ignition sources in different vegetation types as a percentage of total area burned. Values vary from year to year and give only a relative indication of different ignition sources. All the areas were subject to some form of fire management, including both fire suppression and prescribed burning. Data for eucalypt woodlands represent number and not area of fires

Vegetation type	*Region*	*Source*	*Natural (lightning)*	*Human*	*Unknown*
Savanna	South Africa	Trollope (1993)	10	>70	<20
Prairie (pre-20th C.)	North America	Moore (1972)	<1	>99	
Sclerophyll woody	Spain	Vazquez and Moreno (1993)	<10	>60	<30
Chaparral	California	Davis and Michaelson (1995)	<25	75	?
Fynbos	South Africa	van Wilgen (1981)	16	83	1
Fynbos (remote area)	South Africa	Horne (1981)	21	58	21
Eucalypt woodlands	Victoria, Australia	Gill (1981c)	*c.* 20	*c.* 70	10
Boreal forest	North America	Wein and MacLean (1983)	Mostly lightning		

the size of Australia and the largest continuous forest area in the world. The forest is fire-prone (Plates 7 and 8) and 5–10 million ha burn every year (Goldammer, 1993). In the former USSR, the fire management policy was to prevent and control wildfires on protected lands, regardless of the ignition source or the impact of the fire, using hundreds of planes and more than 8000 airborne fire-fighters (Goldammer, 1993). Similar, but smaller, programmes were operated in the United States (Pyne, 1982). The ecological impact of suppression policies on fire regimes and vegetation on this scale seem mind-boggling. However, there is some doubt as to whether fuel suppression ever influenced the incidence of major fires in western American conifer forests (Schullery, 1989).

An insidious and alarming change in fire fuels is also occurring as a result of the invasion of alien plant species in many ecosystems. In chaparral, burnt areas are sometimes reseeded with annual ryegrass, *Lolium*

Table 9.4 The effect of fire on native and alien cover (%) in the seasonal submontane zone of Hawaii. Alien grasses only became common in the area after the late 1960s. Prior to fire, the entire area was an open *Metrosideros polymorpha* (ohia) woodland. From Hughes *et al.* (1991)

Growth form	*Unburnt*	*Burnt, 1987*	*Burnt, 1970*	*Burnt, 1970, 1987*
Native trees	48.7	0.25	0.0	0.0
Native shrubs	63.6	4.9	31.2	0.6
Total native	117.0	5.8	31.6	0.7
Alien grasses	80.0	92.1	101.7	100.9

multiflorum, as an erosion protection measure in the first year after a burn. The grass changes the pattern of fuel accumulation so that fires can occur at exceptionally short frequencies. Zedler *et al.* (1983) reported one instance where ryegrass fuels allowed a fire to spread just 1 year after a natural burn instead of the usual minimum interval of 20 years or more between burns. The effects were devastating, eliminating a non-sprouting *Ceanothus* species and reducing the dominant shrub, chamise, by up to 97% of its pre-burn densities. In Hawaii, introduced grasses have also fostered short rotation fires, eliminating the native Hawaiian forests in just 20 years (Table 9.4; Hughes *et al.*, 1991). In South Africa, invasive pines and Australian *Hakea* have altered fynbos fire regimes, producing exceptionally intense fires which change soil physical properties causing accelerated erosion and flooding in addition to their direct and indirect impact on native species (Scott and van Wyk, 1990; Richardson *et al.*, 1992).

Other ecological processes also affect fuel accumulation and therefore the incidence of fire. In savannas, increased livestock numbers associated with growing human populations reduce the probability of fire through grazing of the grass fuels. Grazing and fire may interact in complex ways to influence vegetation dynamics (Chapter 7).

9.3.3 Changes in the fire landscape

Yet another human impact on past fire regimes is the change in the continuity of fuels. This, perhaps more than any other factor, may disrupt the close climate/fire association seen in the palaeoecological record. The fragmentation of natural habitats has become a major concern over the past few decades and is central to research efforts in conservation biology. Both fragmentation and the aggregation of previously isolated land units influence the pattern of fires. Fragmentation of tropical forests has allowed the incursion of fires into the forests, hastening their demise (Uhl and Kauffman, 1990). On the other hand, fragmentation of fire-dependent vegetation surrounded by non-flammable landscape elements, such as the British heathlands shown in Figure 9.3, may cause fire frequencies to decline. This can result in the local extinction of species which depend on frequent

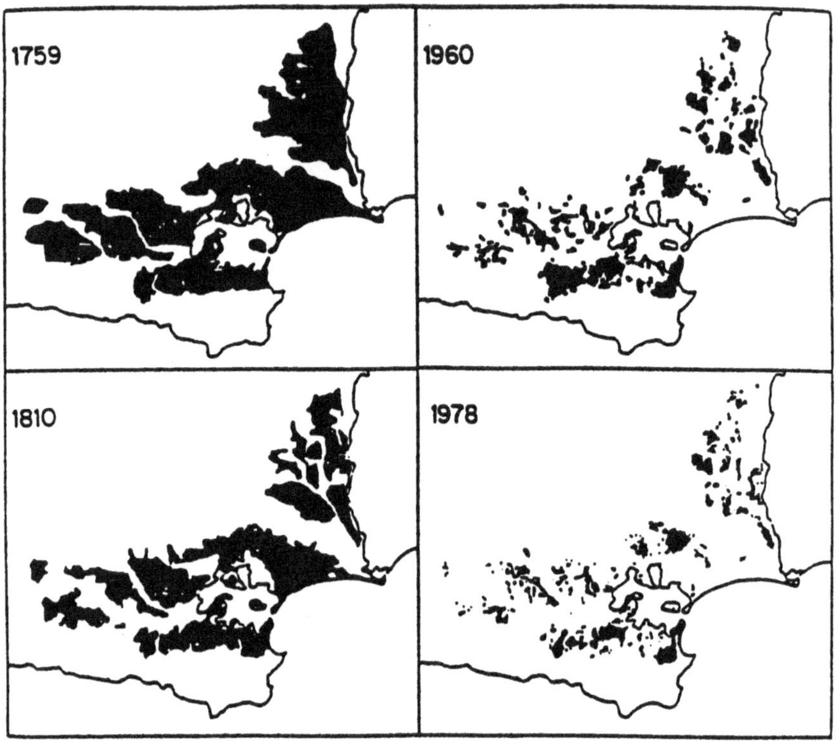

Figure 9.3 Fragmentation of a fire-prone vegetation – heathlands in Dorset, Britain (Webb, 1980). The fire regime is disrupted in such fragmented landscapes, potentially leading to vegetation change.

Table 9.5 Local extinction of species from 'islands' of fynbos shrublands surrounded by non-flammable forests. Extinctions were inferred from differences between plant species lists from islands and matched areas of continuous 'mainland' habitat. Sprouter and short (<1 m) shrub species were significantly under-represented on islands relative to comparable areas of mainland, suggesting that long fire intervals were partly responsible for the losses (Bond *et al.*, 1988)

Area (ha)	*Total species present*		*Island 'extinctions'*	
	Mainland	*Island*	*Mainland – island*	*% loss*
0.5	53	8	45	85
3	70	16	54	77
13	88	26	62	70
25	98	40	58	59
70	115	71	44	38

fires. For example, Bond *et al.* (1988) found far fewer species on islands of fynbos surrounded by non-flammable Afro-temperate forests than in comparable areas of continuous fynbos fuels (Table 9.5). Disproportionately large numbers of sprouting and early successional shrubs were absent on the islands, suggesting that changes in fire regime caused much of the loss. Burning small reserves to maintain fire-dependent species is costly and can promote extinctions because of event-dependent risks of local extinction (Burgman and Lamont, 1992; Chapter 5).

The reverse pattern of aggradation of fire-prone areas is also occurring, especially in industrialized countries. Where fire suppression policies were effective, the landscape now consists of extensive areas of similar-aged fuels with a decline in landscape patchiness (Baker, 1992; Chapter 7). The coalescence of fuels at a landscape scale is also occurring as a result of land abandonment as people flock to urban areas. In southern European countries, this process has led to catastrophic fires (J. Moreno, personal communication, 1994).

The continuity of fuels is also affected by interactions among natural sources of disturbance, including insect outbreaks, vertebrate herbivory or physical processes such as landslides. For example Veblen *et al.* (1994) have recently shown how the extent of a burn in subalpine North American forests depended on changes in fuels associated with snow avalanches and spruce-beetle outbreaks. *Picea* populations did not support a spruce-beetle outbreak until individual trees reached a minimum diameter representing 70 years' growth. The result was non-overlapping distributions between recent fires and beetle outbreaks. The landscape context of disturbance interactions such as these is new in ecology and may reveal unexpected feedbacks to human activity.

Changes in fuel continuity will affect the impact of global warming on fire regimes. In fragmented fuels, it may slow, or even halt vegetation change. Certainly, there will be much greater potential for managing fires to maintain desired communities despite changing climates. The reverse will apply where human activity has increased the continuity of flammable vegetation.

9.4 ECOLOGICAL CONSEQUENCES OF CHANGING FIRE REGIMES

The many human-induced changes in fire regimes pose similar questions to those raised by global climate change:

- How will fire regimes change?
- How will these changes influence small and large-scale patterns of vegetation?
- How quickly will vegetation change in response to changes in fire regimes?
- To what extent can changes be modified or managed?

The physical relationship between fire, climate and fuel is well known at a local scale and is used in a predictive way in fire management (Chapter 2). There is some doubt about the ability of these models to predict fire behaviour accurately under extreme conditions, such as those prevailing during the Yellowstone burn (Rothermel, 1991), though such fires may become more frequent in the future. The potential for developing predictive physical models at larger spatial and temporal scales seems promising from the new research discussed above relating fire, fuel and large-scale climate cycles. These must necessarily consider not only the 'fire behaviour' component of fire propagation but also changes in fuel type and fuel accumulation rates associated with global warming, carbon dioxide increase and vegetation change. Models that simulate fire propagation over landscapes are also becoming available (e.g. Green, 1989; Green *et al.*, 1991).

How will changing fire regimes influence plant assemblages across landscapes? While considerable progress has been made at understanding key fire responses at a species level (Chapters 4 and 5), there have been few attempts to generalize these results to predict responses of whole communities or vegetation formations. Gap models, first developed to simulate forest succession (Botkin *et al.*, 1972; Shugart, 1984), are among the most popular approaches for simulating vegetation dynamics in response to

Box 9.2 Gap models for predicting vegetation dynamics

Botkin *et al.* (1972) developed a model (the JABOWA model) for simulating forest dynamics which has been extended to woody and herbaceous communities throughout the world (Shugart, 1984). The objectives of the original model were to reproduce the dynamic behaviour of a mixed-species, mixed-aged temperate forest and to produce output on the diameter distribution and growth increment of many individual trees. The model simulated the growth of individual trees in a small patch (0.01–0.1 ha). Individual trees had sigmoid growth in diameter under ideal conditions with a species-specific maximum size. The actual growth rate was derived by reducing this ideal rate according to species-specific functions representing the effects of resource availability and climate on growth. Competition was modelled as a change in the light regime, with each individual reducing light to shorter individuals. Shaded individuals 'died' if their increment fell below a threshold. Recruitment was simulated as a lottery drawn from the species list rather than individual trees in the stand, and therefore assumed unlimited dispersal. Shade-tolerant and gap-demanding species had different recruitment requirements. A strength of this class of models is that they are mechanistic. Growth rates can be adjusted according to physiological criteria, including responses to CO_2 increase and global warming.

global warming (Box 9.2). They have now been developed for diverse communities in many parts of the world, including grasslands and savannas (e.g. Prentice *et al.*, 1993). These models simulate the dynamics of a community by modelling the growth of individual trees. They are explicitly reductionist in nature, with the community dynamics depending on the summed responses of individual trees, each growing according to the properties of the species present. Physiological information on tree growth, and likely changes in tree growth due to warming or CO_2 increase, can be incorporated. One reason for the popularity of these models is this capacity to incorporate knowledge of leaf-level phenomena and scale up the consequences for community dynamics.

Gap models were designed to simulate tree growth. In essence they track 'interval-dependent' effects at the level of populations or communities (Chapter 4). How does one incorporate the all-important event-dependent effects on survival and recruitment seen in many fire-prone communities? Gap models have been used in fire-prone environments by incorporating vital attributes to specify fire responses at particular stages of stand development (Shugart and Noble, 1981; Kercher and Axelrod, 1984). Gap models have also been used to simulate fuel dynamics from stand growth. The resultant fuel properties can be used to model changes in fire intensity at different stages of stand development, so that both interval- and intensity-dependent changes in tree survival and recruitment can be simulated (Keane *et al.*, 1990). Large-scale simulations of vegetation dynamics can be generated by deriving very large sets of gap model solutions, each applying to a small plot. Fire is then one of the main factors linking plots or landscape units (Prentice *et al.*, 1993).

It would seem that gap models hold great promise for scaling up vegetation dynamics from site to landscape. But there are problems. The models require a large number of parameters for each species and these are not available for many parts of the world. The response has been to search for 'functional types' (Smith *et al.*, 1993). These are generalized sets of important physiological or life-history characteristics shared by a set of species. The functional type is then used to set the parameters for gap models rather than measures from each individual species. The same need to develop a general set of rules motivated the development of functional types for predicting fire frequency, season and intensity responses (Chapters 4 and 5). Given adequate knowledge of how fire regimes will change, this wider set of functional types could be tested in gap models of vegetation dynamics.

A second problem related to gap models is their complexity. It is difficult to test the sensitivity of these models to changing parameters because of the large number of parameters used. Gap models may provide realistic simulations of vegetation dynamics, but much simpler models may achieve the same results. The dichotomy between event- and interval-dependent philosophies for analysing vegetation dynamics has been emphasized throughout this book. An intensive focus on interval-dependent effects may be unwarranted where disturbance events dominate vegetation dynamics. Why model leaf-level effects on tree growth to predict global warming effects when catastrophic fires are the main agent of change? In a disturbance-driven view of global warming, the fire regime is the driving variable, with life-history attributes providing the best predictor of new vegetation patterns (Chapters 4, 5 and 7). We suggest that models that simulate the fire events, which terminate the intervals simulated by gap models, need much more consideration.

The time taken for stands, and spatial mosaics of stands, to respond to changing climate or fire regimes may be very variable, depending partly on the attributes of individual species. Rates of change may be greatly affected by different degrees of resilience to burning. Maquis shrublands of the Mediterranean, when dominated by obligate sprouters, seem very resilient to fire at different frequencies, seasons and intensities (Trabaud, 1991). Australian and South African sclerophyll shrublands, with a similar physiognomy but dominated by non-sprouters, are highly sensitive to event-dependent effects (Chapter 5). The persistence of species in changing environments may owe much to how individuals persist through disturbance. Sprouting plants with deep-seated buds may persist the longest at a site but, because of poor seedling recruitment, may be slowest to colonize new sites. Resilience to disturbance, whether based on sprouting or seed regeneration, is important for predicting rates of vegetation change due to changing climates or fire regimes. In Mediterranean-type shrublands, many species are likely to persist far longer under changing conditions in the Mediterranean Basin than their southern-hemisphere counterparts.

Inertia in the face of environmental change can also be due to longevity. Many conifers are long-lived and may persist at a site long after climates have become hostile for recruitment. However, old stands may be eliminated abruptly by catastrophic fires where species have no capacity for sprouting. For example, Arseneault and Payette (1992) showed that fire caused a shift from lichen–spruce krummholz to lichen–tundra vegetation in northern Quebec. Forest regeneration failed after a fire in 1750, possibly because of the limited regenerative potential during severe climate conditions of the Little Ice Age. Regeneration failure in turn led to a significant long-term reduction in snow cover in the low post-burn vegetation. The change in snow cover, in turn, triggered frost disturbance of the soil profile by active mudboils, producing long-term changes in the whole ecosystem. Studies such as these suggest that we need to know a great deal more about the resilience of different fire-prone vegetation to extreme fires to understand rates of vegetation change.

9.5 CONCLUSIONS

The central feature of this chapter is change – change in the pressure of human land-use, change in the pattern of landscapes, changing attitudes to fire and its use and changing climates. Changes too, are occurring in research agendas. A common feature of this final chapter is the change in scale of ecological analysis: on the spatial scale from site to landscape and on the temporal scale from years to centuries. Fire is a good vehicle for exploring these new directions in ecology. It is second only to climate in the speed at which it propagates and the spatial extent of its effects. Not surprisingly, fire ecologists have been among the first to explore the role of disturbance at large scales and over long time periods. However, there are still major problems to be overcome in extrapolating population level processes to larger scales. There is increasingly sophisticated understanding of the basis for fire responses at a population level (Chapter 5) but there have been few attempts to translate this to the distribution of species or communities across landscapes (Chapter 7). Fire behaviour can be very heterogeneous across small distances and in similar fuels burnt at different times. New analytical methods need to be developed to explore the implications of this heterogeneity. For large-scale predictions of fire effects, new rules will have to be found, more general than those based on detailed population studies. A key prerequisite for research progress will be a clear understanding of the relevant scale for which answers are required when designing new research projects.

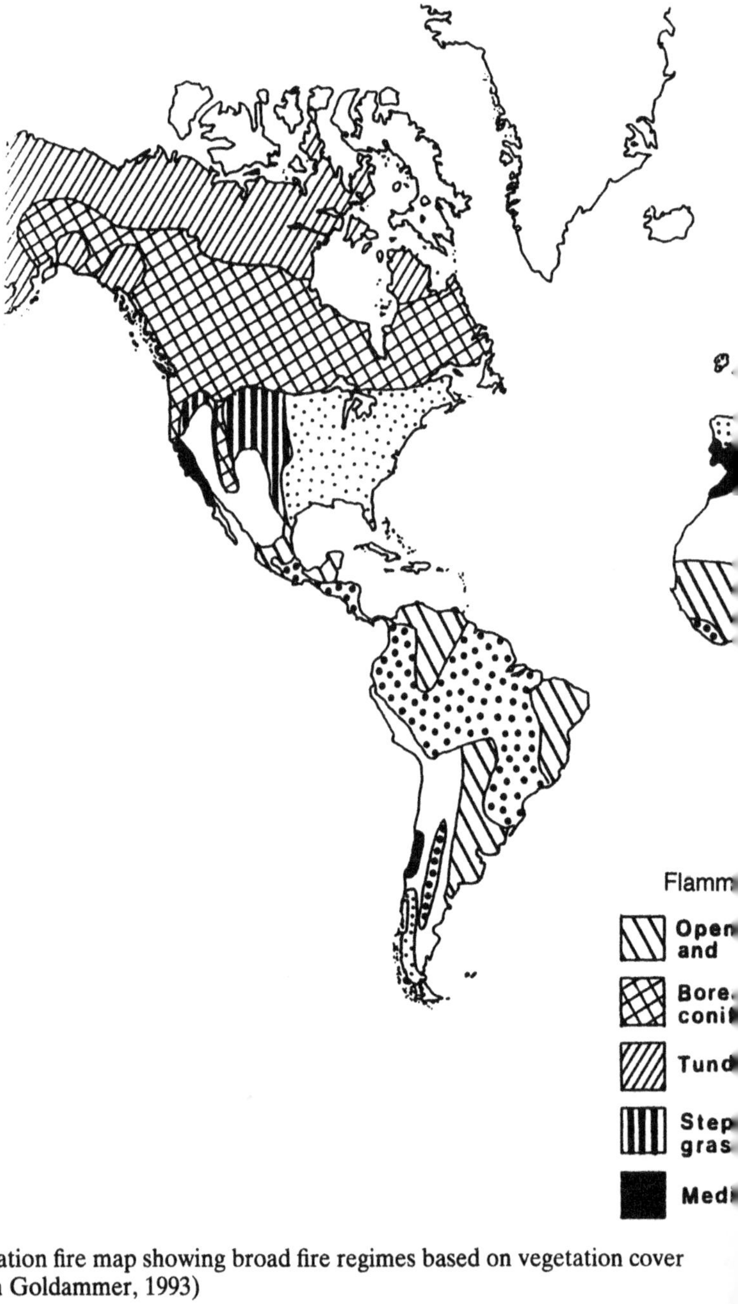

A global vegetation fire map showing broad fire regimes based on vegetation cover (modified from Goldammer, 1993)

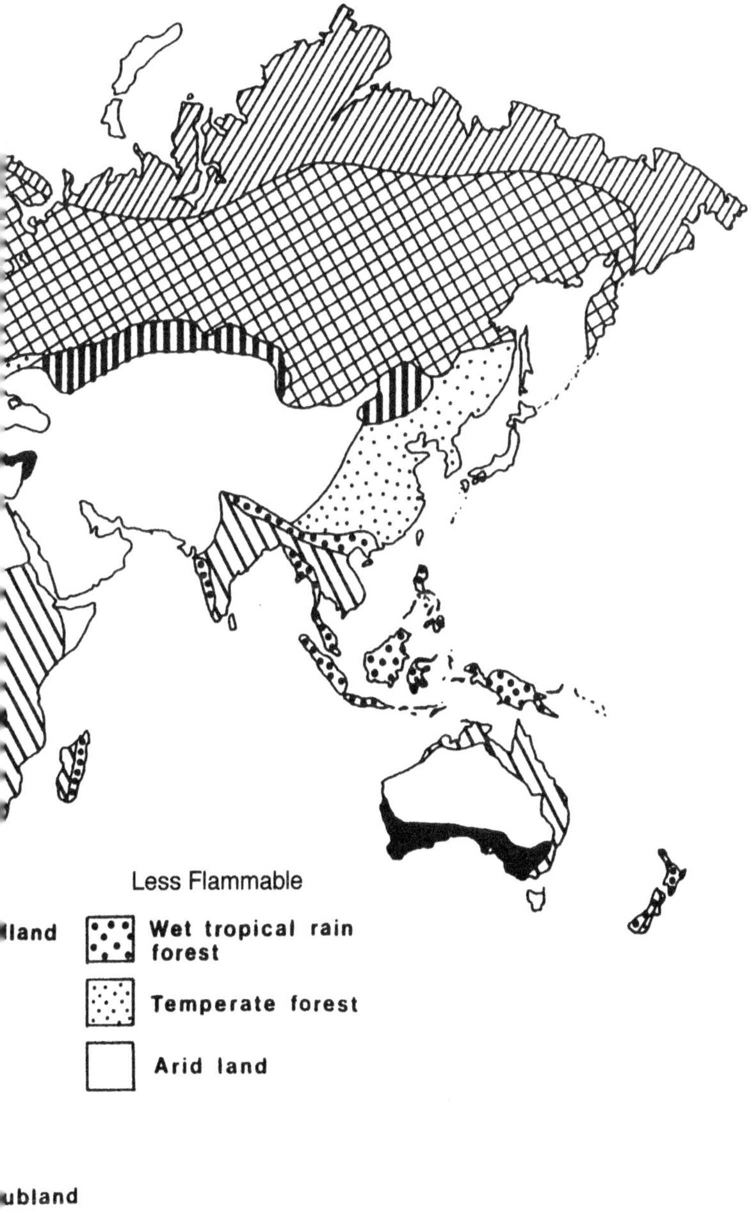
Less Flammable
lland
Wet tropical rain forest
Temperate forest
Arid land
ubland

References

Abrams, P.A. (1993) Effect of increased productivity on the abundance of trophic levels. *American Naturalist*, **141**, 351–71.

Alexander, M.E. (1982) Calculating and interpreting forest fire intensities. *Canadian Journal of Botany*, **60**, 349–57.

Andrasko, K.J., Ahuja, D.R., Winnett, S.M. and Tirpak, D.A. (1991) Policy options for managing biomass burning to mitigate global climate change, in *Global Biomass Burning: Atmospheric, Climatic and Biospheric Implications*, (ed. J.S. Levine), MIT Press, Cambridge, MA, pp. 445–56.

Andrews, P.L. (1986) BEHAVE: Fire Behaviour Prediction and Fuel Modelling System – BURN Subsystem. *USDA Forest Service General Technical Report INT-194.*

Andrews, P.L. and Chase, C.H. (1989) BEHAVE: Fire Behaviour and Fuel Modelling System – BURN Subsystem, Part 2. *USDA Forest Service General Technical Report INT-260.*

Archer, S., Scifres, C., Bassham, C.R. and Maggio, R. (1988) Autogenic succession in a subtropical savanna: conversion of grassland to thorn woodland. *Ecological Monographs*, **58**, 111–27.

Arseneault, D. and Payette, S. (1992) A postfire shift from lichen–spruce to lichen–tundra vegetation at tree line. *Ecology*, **73**, 1067–81.

Ash, J. (1988) The location and stability of rainforest boundaries in north-eastern Queensland, Australia. *Journal of Biogeography*, **15**, 619–30.

Ashton, D.H. (1981) Fire in tall open forests (wet sclerophyll forests), in *Fire and the Australian Biota,* (eds A.M. Gill, R.H. Groves and I.R. Noble), Australian Academy of Science, Canberra, pp. 339–66.

Aston, A.R. and Gill, A.M. (1976) Coupled soil moisture, heat and water vapour transfers under simulated fire conditions. *Australian Journal of Soil Research*, **14**, 55–66.

Auerbach, M. and Shmida, A. (1987) Spatial scale and the determinants of plant species richness. *Trends in Ecology and Evolution*, **2**, 238–42.

Augspurger, C.K. (1984) Reproductive synchrony of a tropical shrub: experimental studies on effects of pollinators and seed predators on *Hybanthus prunifolius* (Violaceae). *Ecology*, **62**, 775–88.

Auld, T.D. (1986) Population dynamics of the shrub *Acacia suaveolens* (Sm.) Willd: fire and the transition to seedlings. *Australian Journal of Ecology*, **11**, 373–85.

Auld, T.D. (1987) Population dynamics of the shrub *Acacia suaveolens* (Sm.) Willd: survivorship throughout the life cycle, a synthesis. *Australian Journal of Ecology*, **12**, 139–52.

Auld, T.D. (1990) The survival of juvenile plants of the resprouting shrub *Angophora hispida* (Myrtaceae) after a simulated low-intensity fire. *Australian Journal of Botany*, **38**, 255–60.

Auld, T.D. and O'Connell, M.A. (1991) Predicting patterns of post-fire germination in 35 eastern Australian Fabaceae. *Australian Journal of Ecology*, **16**, 53–70.

Axelrod, D.I. (1985) Rise of the grassland biome, Central North America. *Botanical Review*, **51**, 163–201.

Backeus, I. (1992) Distribution and vegetation dynamics of humid savannas in Africa and Asia. *Journal of Vegetation Science*, **3**, 345–56.

Baker, W.L. (1992) Effects of settlement and fire suppression on landscape structure. *Ecology*, **73**, 1879–87.

Ball, G.L. and Guertin, D.P. (1992) Improved fire growth modelling. *International Journal of Wildland Fire*, **2**, 47–54.

Ballardie, R.T. and Whelan, R.J. (1986) Masting, seed dispersal and seed predation in the cycad *Macrozamia communis*. *Oecologia*, **60**, 100–5.

Baxter, B.J.M., van Staden, J., Granger, J.E. and Brown, N.A.C. (1994) Plant-derived smoke and smoke extracts stimulate seed germination of the fire-climax grass *Themeda triandra*. *Environmental and Experimental Botany*, **34**, 217–23.

Bazzaz, F.A. (1979) The physiological ecology of plant succession. *Annual Review of Ecology and Systematics*, **56**, 485–8.

Beeby, A. (1993) *Applying Ecology*, Chapman & Hall, London, p. 441.

Begon, M., Harper, J.L. and Townsend, C.R. (1990) *Ecology: Individuals, Populations and Communities*, Blackwell, Oxford.

Bell, D.T., Hopkins, A.J.M. and Pate, J.S. (1984) Fire in the kwongan, in *Kwongan: Plant Life of the Sandplain*, (eds J.S. Pate and J.S. Beard), University of Western Australia Press, Nedlands, pp. 178–204.

Bell, D.T., Plummer, J.A. and Taylor, S.K. (1993) Seed germination ecology in Southwestern Western Australia. *Botanical Review*, **59**, 24–73.

Bellairs, J.M. and Bell, D.T. (1990) Canopy-borne seed store in three Western Australian plant communities. *Australian Journal of Ecology*, **15**, 299–305.

Bellows, T.S. (1981) The descriptive properties of some models for density dependence. *Journal of Animal Ecology*, **50**, 139–56.

Belsky, A.J. (1986) Does herbivory benefit plants? A review of the evidence. *American Naturalist*, **127**, 870–92.

Belsky, A.J. (1990) Tree/grass ratios in East African savannas: a comparison of existing models. *Journal of Biogeography*, **17**, 483–9.

Berg, A.R. and Plumb, T.R. (1972) Bud activation for regrowth, in *Wildland Shrubs – Their Biology and Utilization*, (eds C.M. McKell, J.P. Blaisdell and J.R. Goodin), USDA For. Serv. Intermount. Forest and Range Exp. Sta. Gen.Tech. Rep. INT-1. Ogden, Utah, pp. 279–86.

Bergeron, Y. (1991) The influence of island and mainland landscapes on the boreal forest fire regime. *Ecology*, **72**, 1980–92.

Bergeron, Y. and Brisson, J. (1990) Fire regime in red pine stands at the northern limit of the species range. *Ecology*, **71**, 1352–64.

Berryman, A.A. (1991) Stabilization or regulation: what it all means! *Oecologia*, **86**, 140–3.

Berryman, A.A. and Milstein, J.A. (1989) Are ecological systems chaotic and if not, why not? *Trends in Ecology and Evolution*, **4**, 26–8.

Bertness, M.D. (1991) Interspecific interactions among high marsh perennials in a New England salt marsh. *Ecology*, **72**, 125–37.

Bidwell, T.G., Engle, D.M. and Claypool, P.L. (1990) Effects of spring headfires and backfires on tallgrass prairie. *Journal of Range Management*, **43**, 209–12.

Bigalke, R.C. and Willan, K. (1984) Effects of fire regime on faunal composition and dynamics, in *Ecological Effects of Fire in South African Ecosystems*, (eds P. de V. Booysen and N.M. Tainton), Springer-Verlag, Berlin, pp. 255–72.

Biswell, H.H. (1989) *Prescribed Burning in California Wildlands Management*, University of California Press, Berkeley.

Bond, W.J. (1980) Fire and senescent fynbos in the Swartberg, Southern Cape. *South African Forestry Journal*, **114**, 68–71.

Bond, W.J. (1983) Dead leaves and fire survival in Southern African tree aloes. *Oecologia*, **58**, 110–14.

Bond, W.J. (1984) Fire survival of Cape Proteaceae – influence of fire season and seed predators. *Vegetatio*, **56**, 65–74.

Bond, W.J. (1985) Canopy stored seed reserves (serotiny) in Cape Proteaceae. *South African Journal of Botany*, **51**, 181–6.

Bond, W.J. (1989) Describing and conserving biotic diversity, in *Biotic Diversity in Southern Africa*, (ed. B.J. Huntley), Oxford University Press, Cape Town, pp. 2–18.

Bond, W.J. and Midgley, J.J. (1995) Kill thy neighbour: an individualistic argument for the evolution of flammability. *Oikos*, **73**, 79–85.

Bond, W.J., Vlok, J. and Viviers, M. (1984) Variation in seedling recruitment of Cape Proteaceae after fire. *Journal of Ecology*, **72**, 209–21.

Bond, W.J., Midgley, J.J. and Vlok, J. (1988) When is an island not an island? Insular effects and their causes in fynbos shrublands. *Oecologia*, **77**, 515–21.

Bond, W.J., Le Roux, D. and Erntzen, R. (1990) Fire intensity and regeneration of myrmecochorous Proteaceae. *South African Journal of Botany*, **56**, 326–31.

Bond, W.J., Cowling, R.M. and Richards, M.B. (1992). Competition and coexistence in proteoid shrubs, in *The Ecology of Fynbos: Nutrients, Fire and Diversity*, (ed. R.M. Cowling), Oxford University Press, Cape Town, pp. 206–25.

Booysen, P. de V. and Tainton, N.M. (eds) (1984) *Ecological Effects of Fire in South African Ecosystems*, Ecological Studies 48, Springer-Verlag, Berlin.

Booysen, P. de V., Tainton, N.M. and Scott, J.D. (1963) Shoot apex development in grasses and its importance in grassland management. *Herbage Abstracts*, **33**, 209–13.

Borchert, M. (1985) Serotiny and cone-habit variation in populations of *Pinus coulteri* (Pinaceae) in the southern coast ranges of California. *Madrono*, **32**, 29–48.

Bosch, J.M. and Hewlett, J.D. (1982) A review of catchment experiments to determine the effect of vegetation changes on water yield and evapotranspiration. *Journal of Hydrology*, **55**, 3–23.

Botkin, D.B., Janak, J.F. and Wallis, J.R. (1972) Some ecological consequences of a computer model of forest growth. *Journal of Ecology*, **60**, 849–73.

Boucher, C. (1981) Autecological and population studies of *Orothamnus zeyheri* in the Cape of South Africa, in *The Biological Aspects of Rare Plant Conservation*, (ed. H. Synge), Wiley, New York, pp. 343–53.

Bowman, D.M.J.S. and Wilson, B.A. (1988) Fuel characteristics of coastal monsoon forest, Northern Territories, Australia. *Journal of Biogeography*, **15**, 807–17.

Bowman, D.M.J.S., Wilson, B.A. and Hooper, R.J. (1988) Response of *Eucalyptus* forest and woodland to four fire regimes at Munmarlary, Northern Territory, Australia. *Journal of Ecology*, **76**, 215–32.

Bradstock, R.A. (1990) Demography of woody plants in relation to fire: *Banksia serrata* Lf. and *Isopogon anemonifolius* (Salisb.) Knight. *Australian Journal of Ecology*, **15**, 117–32.

Bradstock, R.A. and Bedward, M. (1992) Simulation of the effect of season of fire on post-fire seedling emergence of two *Banksia* species based on long-term rainfall records. *Australian Journal of Botany*, **40**, 75–88.

Bradstock, R.A. and O'Connell, M.A. (1988) Demography of woody plants in relation to fire: *Banksia ericifolia* L.f. and *Petrophile pulchella* (Schrad) R.Br. *Australian Journal of Ecology*, **13**, 505–18.

Brain, C.K. and Sillen, A. (1988) Evidence from the Swartkrans cave for the earliest use of fire. *Nature*, **336**, 464–6.

Brenner, J. (1991) Southern Oscillation anomalies and their relationship to wildfire activity in Florida. *International Journal of Wildland Fire*, **1**, 73–8.

Brits, G.J., Calitz, F.J., Brown, N.A.C. and Manning, J.C. (1993) Desiccation as the active principle in heat-stimulated seed germination of *Leucospermum* R.Br. (Proteaceae) in fynbos. *New Phytologist*, **125**, 397–403.

Brookman-Amissah, J., Hall, J.B., Swaine, M.D. and Attakorah, J.Y. (1980) A re-assessment of a fire protection experiment in north-eastern Ghana savanna. *Journal of Applied Ecology*, **17**, 85–99.

Brown, N.A.C. (1993) Promotion of germination of fynbos seeds by plant-derived smoke. *New Phytologist*, **122**, 1–9.

Brown, P.J., Manders, P.T., Bands, D.P. *et al.* (1991) Prescribed burning as a conservation management practice: a case history from the Cederberg mountains, Cape Province, South Africa. *Biological Conservation*, **56**, 133–50.

Buckley, R. (1984) The role of fire in maintaining a small-scale vegetation gradient: a test of Mutch's hypothesis. *Oikos*, **41**, 291–2.

Buckman, R.E. (1964) Effects of a prescribed burn on hazel in Minnesota. *Ecology*, **45**, 626–9.

Burgan, R.E. and Rothermel, R.C. (1984) BEHAVE: Fire Behaviour Prediction and Fuel Modelling System. Fuel Subsystem. *USDA Forest Service General Technical Report INT-167.*

Burgman, M.A. and Lamont, B.B. (1992) A stochastic model for the viability of *Banksia cuneata* populations: environmental, demographic and genetic effects. *Journal of Applied Ecology*, **29**, 719–27.

Burney, D.A. (1987) Pre-settlement vegetation changes at Lake Tritrivakely, Madagascar. *Paleoecology of Africa*, **18**, 357–81.

Busso, C.A., Boo, R.M. and Pelaez, D.V. (1993) Fire effects on bud viability and growth of *Stipa tenuis* in semi-arid Argentina. *Annals of Botany*, **71**, 377–81.

Byram, G.M. (1959) Combustion of forest fuels, in *Forest Fire: Control and Use*, (ed. K.P. Davis), McGraw-Hill, New York, pp. 155–82.

Byram, G.M. (1963) An analysis of the drying process in forest fuel material. Paper presented at the 1960 International Symposium on Humidity and Moisture, Washington DC.

Caldwell, M.M. (1986) Ecophysiology of rangeland plants, in *Rangelands: A Resource under Siege*, (eds P.J. Joss, P.W. Lynch and O.B. Williams), Cambridge University Press, Cambridge, pp. 423–4.

Carpenter, F.L. and Recher, H.F. (1979) Pollination, reproduction and fire. *American Naturalist*, **113**, 871–9.

Carr, D.J., Jahnke, R. and Carr, S.G.M. (1984) Initiation, development and anatomy of lignotubers in some species of *Eucalyptus*. *Australian Journal of Botany*, **32**, 415–37.

Caswell, H. (1982) Optimal life histories and the maximization of reproductive value: a general theorem for complex life cycles. *Ecology*, **63**, 1218–22.

Caswell, H. (1989) *Matrix Population Models: Construction, Analysis, and Interpretation*, Sinauer Associates, Sunderland, MA.

Cavanagh, A.K. (1980) A review of some aspects of the germination of Acacias. *Proceedings of the Royal Society of Victoria*, **91**, 161–80.

Chandler, C., Cheney, P., Thomas, P. *et al.* (1983) *Fire in Forestry*, vol. II *Forest Fire Management and Organisation*, John Wiley and Sons, New York.

Charnov, E.L. and Shaffer, W.M. (1973) Life history consequences of natural selection: Cole's result revisited. *American Naturalist*, **117**, 923–43.

Cheney, N.P. (1981) Fire behaviour, in *Fire and the Australian Biota*, (eds A.M. Gill, R.H. Groves and I.R. Noble), Australian Academy of Science, Canberra, pp. 151–76.

Chesson, P.L. and Warner, R.R. (1981) Environmental variability promotes coexistence in lottery competitive systems. *American Naturalist*, **117**, 923–43.

Christensen, N.L. and Muller, C.H. (1975) Effects of fire on factors controlling plant growth in *Adenostoma* chaparral. *Ecological Monographs*, **45**, 29–55.

Clark, J.S. (1988) Effects of climate change on fire regime in north-western Minnesota. *Nature*, **334**, 233–5.

Clark, J.S. (1989) Ecological disturbance as a renewal process: theory and application to fire history. *Oikos*, **56**, 17–30.

Clark, J.S. (1990) Fire and climate change during the last 750yr in northwestern Minnesota. *Ecological Monographs*, **60**, 135–59.

Clark, J.S. (1991) Disturbance and tree life history on the shifting mosaic landscape. *Ecology*, **72**, 1102–18.

Clement, B. and Touffet, J. (1981) Vegetation dynamics in Brittany heathlands after fire. *Vegetatio*, **46**, 157–66.

Clement, B. and Touffet, J. (1990) Plant strategies and secondary succession on Brittany heathlands after severe fire. *Journal of Vegetation Science*, **1**, 195–202.

Clements, F.E. (1916) *Plant Succession*, Carnegie Institute, Washington, publ. 242.

Cody, M.L. (1986) Structural niches in plant communities, in *Community Ecology*, (eds J. Diamond and T. Case), Harper and Row, New York, pp. 381–405.

Cody, M.L. and Mooney, H.A. (1978) Convergence versus nonconvergence in Mediterranean-climate ecosystems. *Annual Review of Ecology and Systematics*, **9**, 265–321.

Coley, P.D. (1988) Effects of plant growth rate and leaf life time on the amount and type of antiherbivore defense. *Oecologia*, **74**, 531–6.

Coley, P.D., Bryant, J.P. and Stuart Chapin III, F. (1985) Resource availability and plant antiherbivore defense. *Science*, **230**, 895–9.

Colinvaux, P.A. (1986) *Ecology*, Wiley, New York.

Collins, S.L. (1992) Heterogeneity of disturbed tallgrass prairie communities. *Ecology*, **73**, 2001–6.

Collins, S.L. and Gibson, D.J. (1990) Effects of fire on community structure in tall-grass and mixed-grass prairie, in *Fire in North American Tallgrass Prairies*, (eds S.L. Collins and L.L. Wallace), University of Oklahoma Press, Norman, pp. 81–98.

Connell, J.H. (1978) Diversity in tropical rainforests and coral reefs. *Science*, **199**, 1302–10.

Connell, J.H. and Slatyer, R.O. (1977) Mechanisms of succession in natural communities and their role in community stability and organisation. *American Naturalist*, **111**, 1119–44.

Countryman, C.M. and Philpot, C.W. (1970) Physical characteristics of chamise as a wildland fuel. *USDA Forest Service Research Paper PSW 66*, Berkeley, California.

Coutinho, L.M. (1982) Ecological effects of fire in Brazilian Cerrado, in *Ecology of Tropical Savannas*, (eds B.J. Huntley and B.H. Walker), Ecological Studies 42, Springer-Verlag, Berlin, pp. 273–91.

Cowling, R.M. (1987) Fire and its role in coexistence and speciation in Gondwana shrublands. *South African Journal of Science*, **83**, 106–12.

Cowling, R.M. and Gxaba, T. (1990) Effects of a fynbos overstorey shrub on understorey community structure: implications for the maintenance of community-wide species richness. *South African Journal of Ecology*, **1**, 1–7.

Cowling, R.M. and Lamont, B.B. (1985a) Variation in serotiny of three *Banksia* species along a climatic gradient. *Australian Journal of Ecology*, **10**, 345–50.

Cowling, R.M. and Lamont, B.B. (1985b) Seed release in *Banksia*: the role of wet–dry cycles. *Australian Journal of Ecology*, **10**, 169–71.

Cowling, R.M. and Lamont, B.B. (1987) Post-fire recruitment of four co-occurring *Banksia* species. *Journal of Applied Ecology*, **24**, 645–58.

Cowling, R.M., Lamont, B.B. and Enright, N.J. (1990) Fire and management of south-western Australian banksias. *Proceedings of the Ecological Society of Australia*, **16**, 177–83.

Crawley, M.J. (1983) *Herbivory: the Dynamics of Animal-Plant Interactions*, Blackwell, Oxford.

Crawley, M.J. (1986) The structure of plant communities, in *Plant Ecology*, (ed. M.J. Crawley), Blackwell, Oxford, pp. 1–50.

Crawley, M.J. (1990) Plant population dynamics. *Philosophical Transactions of the Royal Society, Series B*, **314**, 125–40.

Crutzen, P.J. and Andreae, M.O. (1990) Biomass burning in the tropics: impact on atmospheric chemistry and biogeochemical cycles. *Science*, **250**, 1669–78.

D'Antonio, C.M. and Vitousek, P.M. (1992) Biological invasions by exotic grasses, the grass/fire cycle, and global change. *Annual Review of Ecology and Systematics*, **23**, 63–87.

Daubenmire, R.F. (1968) Ecology of fire in grasslands. *Advances in Ecological Research*, **5**, 203–26.

Daubenmire, R.F. (1974) *Plants and Environment: A Textbook of Plant Autecology*, Wiley, New York.

Davis, F.W. and Michaelson, J. (1995) Sensitivity of fire regime in chaparral ecosystems to global climate change, in *Global Climate Change in Mediterranean-Type Ecosystems*, (eds J.M. Moreno and W.C. Oechel), Ecological Studies, Springer-Verlag, Berlin (in press).

Davis, S.D. (1991) Lack of niche differentiation in adult shrubs implicates the importance of the regeneration niche. *Trends in Ecology and Evolution*, **6**, 272–4.

Davis, S.D. and Mooney, H.A. (1986) Water use patterns of four co-occurring chaparral shrubs. *Oecologia*, **70**, 172–7.

DeAngelis, D.L. and Waterhouse, J.C. (1987) Equilibrium and nonequilibrium concepts in ecological models. *Ecological Monographs*, **57**, 1–21.

DeBano, L.F., Rice, R.M. and Conrad, C.E. (1979) Soil heating in chaparral fires: effects on soil properties, plant nutrients, erosion, and run-off. *USDA Forest Service Research Paper PSW-145*, Berkeley, California.

Deeming, J.E. and Brown, J.K. (1975) Fuel models in the National Fire Danger Rating System. *Journal of Forestry*, **73**, 347–80.

Deevey, E.S. (1947) Life tables for natural populations of animals. *Quarterly Review of Biology*, **22**, 283–314.

De Lange, J.H. and Boucher, C. (1990) Autecological studies on *Audouinia capitata* (Bruniaceae). I. Plant-derived smoke as a germination cue. *South African Journal of Botany*, **56**, 700–3.

De Lange, J.H. and Boucher, C. (1992) Autecological studies on *Audouinia capitata* (Bruniaceae). 8. Role of fire in regeneration. *South African Journal of Botany*, **59**, 188–202.

Dell, J.D. and Philpot, C.W. (1965) Variations in the moisture content of several fuel size components of live and dead chamise. *USDA Forest Service Research Note PSW-83*.

Dickinson, K.J.M. and Kirkpatrick, J.B. (1985) The flammability and energy content of some important plant species and fuel components in the forests of south-eastern Tasmania. *Journal of Biogeography*, **12**, 121–34.

Dixon, K.W., Roche, S. and Pate, J.S. (1995) The promotive effect of smoke derived from burnt native vegetation on seed germination of Western Australian plants. *Oecologia*, **101**, 185–92.

Driscoll, R.S. (1963) Sprouting bitterbrush in Central Oregon. *Ecology*, **44**, 820–1.

Dublin, H.T., Sinclair, A.R.E. and McGlade, J. (1990) Elephants and fire as causes of multiple stable states in the Serengeti-Mara woodlands. *Journal of Animal Ecology*, **59**, 1147–64.

Edwards, P.J. (1989) Insect herbivory and plant defence theory, in *Toward a More Exact Ecology*, (eds P.J. Grubb and J.B. Whittaker), Blackwell, Oxford, pp. 275–98.

Egler, F.E. (1954) Vegetation science concepts, I. Initial floristic composition – a factor in old-field vegetation development. *Vegetatio*, **4**, 412–17.

Ehrlich, P.R. and Roughgarden, J. (1987) *The Science of Ecology*, Macmillan, New York.

Ellison, L. (1960) Influence of grazing on plant succession of rangelands. *Botanical Reviews*, **26**, 1–78.

Enright, N.J. and Lamont, B.B. (1989) Seed banks, fire season, safe sites and seedling recruitment in five co-occurring *Banksia* species. *Journal of Ecology*, **77**, 1111–22.

Everson, C.S., Everson, T.M. and Tainton, N.M. (1985) The dynamics of *Themeda triandra* tillers in relation to burning in the Natal Drakensberg. *Journal of the Grassland Society of Southern Africa*, **2**, 18–25.

Everson, C.S., Everson, T.M. and Tainton, N.M. (1988) Effects of intensity and height of shading on the tiller initiation of six grass species from the Highland sourveld of Natal. *South African Journal of Botany*, **54**, 315–18.

Fagerstrom, T. (1988) Lotteries in communities of sessile organisms. *Trends in Ecology and Evolution*, **3**, 303–6.

Feely, J.M. (1980) Did Iron Age Man have a role in the history of Zululand's wilderness landscapes? *South African Journal of Science*, **76**, 150–2.

Flinn, R.C., Scifres, C.J. and Archer, S.R. (1992) Variation in basal sprouting in co-occurring shrubs: implications for stand dynamics. *Journal of Vegetation Science*, **3**, 125–8.

Fox, B.J., Quinn, R.D. and Breytenbach, G.J. (1985) A comparison of small-mammal succession following fire in shrublands of Australia, California and South Africa. *Proceedings of the Ecological Society of Australia*, **14**, 179–97.

Fretwell, S.D. (1987) Food chain dynamics: the central theory of ecology? *Oikos*, **50**, 291–301.

Frost, P.G.H. (1984) The responses and survival of organisms in fire-prone environments, in *Ecological Effects of Fire in South African Ecosystems*, (eds P. de V. Booysen and N.M. Tainton), Springer Verlag, Berlin, pp. 273–309.

Frost, P.G., Medina, E., Menaut, J.C. *et al.* (1986) Responses of savannas to stress and disturbance. *Biology International*, Special Issue 10.

Gadgil, M. and Solbrig, O.T. (1972) The concept of r- and K- selection: evidence from wild flowers and some theoretical considerations. *American Naturalist*, **106**, 14–31.

Geldenhuys, C.J. (1977) The effect of different regimes of annual burning on two woodland communities in Kavango. *South African Forestry Journal*, **103**, 32–42.

Geldenhuys, C.J. (1994) Bergwind fires and the location pattern of forest patches in the southern Cape landscape, South Africa. *Journal of Biogeography*, **21**, 49–62.

Gill, A.M. (1976) Fire and the opening of *Banksia ornata* F. Muell. follicles. *Australian Journal of Botany*, **24**, 329–35.

Gill, A.M. (1975) Fire and the Australian Flora: a review. *Australian Forestry*, **38**, 4–25.

Gill, A.M. (1981a) Adaptive responses of Australian vascular plant species to fires, in *Fire and The Australian Biota*, (eds A.M. Gill, R.H. Groves and I.R. Noble), Australian Academy of Science, Canberra, pp. 243–72.

Gill, A.M. (1981b) Fire adaptive traits of vascular plants, in Proceedings of the Conference on Fire Regimes and Ecosystem Properties, (eds H.A. Mooney, T.M. Bonnicksen, N.L. Christensen, J.E. Lotan and W.A. Reiners), *USDA Forest Service, General Technical Report WO-26*, pp. 208–30.

Gill, A.M. (1981c) Post-settlement fire history in Victorian Landscapes, in *Fire and The Australian Biota*, (eds A.M. Gill, R.H. Groves and I.R. Noble), Australian Academy of Sciences, Canberra, pp. 77–100.

Gill, A.M. and Ingwersen, F. (1976) Growth of *Xanthorrhoea australis* R.Br. in relation to fire. *Journal of Applied Ecology*, **13**, 195–203.

Gill, A.M. and McMahon, A. (1986) A post-fire chronosequence of cone, follicle and seed production in *Banksia ornata*. *Australian Journal of Botany*, **34**, 425–33.

Gill, A.M., Hoare, J.R.L. and Cheney, N.P. (1990) Fires and their effects in the wet–dry tropics of Australia, in *Fire in the Tropical Biota*, (ed. J.G. Goldammer), Ecological Studies 84, Springer-Verlag, Berlin, pp. 159–78.

Gillon, D. (1983) The fire problem in tropical savannas, in *Tropical Savannas*, (ed. F. Bourliere), Elsevier, Amsterdam, pp. 617–41.

Gimingham, C.H. (1988) A reappraisal of cyclical processes in *Calluna* heath. *Vegetatio*, **77**, 61–4.

Givnish, T.J. (1981) Serotiny, geography and fire in the pine barrens of New Jersey. *Evolution*, **35**, 101–23.

Glenn, S.M., Collins, S.L. and Gibson, D.J. (1992) Disturbances in tallgrass prairie: local and regional effects on community heterogeneity. *Landscape Ecology*, **7**, 243–51.

Glenn-Lewin, D.C., Johnson, L.A., Jurik, T.W. *et al.* (1990) Fire in central North American grasslands: vegetative reproduction, seed germination, and seedling establishment, in *Fire in North American Tallgrass Prairies*, (eds S.L. Collins and L.L. Wallace), University of Oklahoma Press, Norman, pp. 28–45.

Glenn-Lewin, D.C., Peet, R.K. and Veblen, T.T. (1992) *Plant Succession: Theory and Prediction*, Chapman & Hall, London.

Goldammer, J.G. (1993) Wildfire management in forests and other vegetation: a global perspective. *Disaster Management*, **5**, 3–10.

Goldammer, J.G. and Seibert, B. (1990) The impacts of droughts and forest fires on tropical lowland rain forests of Eastern Borneo, in *Fire in the Tropical Biota*, (ed. J.G. Goldammer), Ecological Studies 84, Springer-Verlag, Berlin, pp. 11–31.

Grace, J. and Tilman, D. (eds) (1990) Preface, in *Perspectives on Plant Competition*, Academic Press, New York.

Green, D.G. (1989) Simulated effects of fire, dispersal and spatial pattern on competition within forest mosaics. *Vegetatio*, **82**, 139–53.

Green, D.G., Tridgell, A. and Gill, A.M. (1991) Interactive simulation of bushfires in heterogeneous fuels. *Mathematical and Computer Modelling*, **13**, 57–66.

Green, G.M. and Sussman, R.W. (1990) Deforestation history of the eastern rain forests of Madagascar from satellite images. *Science*, **248**, 212–15.

Green, L.R. (1981) Burning by prescription in chaparral. *USDA Forest Service, General Technical Report PSW-51.*

Grime, J.P. (1979) *Plant Strategies and Vegetation Processes*. John Wiley, Chichester.

Grimm, E.C. (1984) Fire and other factors controlling the big woods vegetation of Minnesota in the mid-nineteenth century. *Ecological Monographs*, **54**, 291–311.

Grubb, P. (1977) The maintenance of species richness in plant communities: the importance of the regeneration niche. *Biological Reviews*, **52**, 107–45.

Gunster, A. (1992) Aerial seedbanks of the central Namib: Distribution of serotinous species in relation to climate and habitat. *Journal of Biogeography*, **19**, 563–72.

Hairston, N.G., Smith, F.E. and Slobodkin, L.B. (1960) Community structure, population control, and competition. *American Naturalist*, **94**, 421–5.

Hamrick, J.L. and Godt, M.J.W. (1990) Allozyme diversity in plant species, in *Plant Population Genetics, Breeding and Genetic Resources*, (eds A.H.D. Brown, M.T. Clegg, A.L. Kahler and B.S. Weir), Sinauer, Sunderland, pp. 43–63.

Hanes, T.L. (1965) Ecological studies on two closely related chaparral shrubs in southern California. *Ecological Monographs*, **35**, 213–35.

Hanes, T.L. (1971) Succession after fire in the chaparral of southern California. *Ecological Monographs*, **41**, 27–52.

Hansen, A.J., Spies, T.A. and Swanson, F.J. (1991a) Conserving biodiversity in managed forests: lessons from natural forests. *Bioscience*, **41**, 382–4.

Hansen, A.J., Pate, J.S. and Hansen, A.P. (1991b) Growth and reproductive performance of a seeder and a resprouter species of *Bossaiea* as a function of plant age after fire. *Annals of Botany*, **67**, 497–509.

Hare, R.C. (1965) Contribution of bark to fire resistance in southern trees. *Journal of Forestry*, **63**, 248–51.

Harper, J.L. (1977) *The Population Biology of Plants*, Academic Press, London.

Harper, J.L. and White, J. (1974) The demography of plants. *Annual Review of Ecology and Systematics*, **5**, 419–63.

Harrington, M.G. (1985) The effects of spring, summer, and fall burning on Gambel Oak in a Southwestern Ponderosa pine stand. *Forest Science*, **31**, 156–63.

Harrington, M.G. (1989) Gambel oak root carbohydrate response to spring, summer, and fall prescribed burning. *Journal of Range Management*, **42**, 504–7.

Heinselman, M.L. (1973) Fire in the virgin forests of the Boundary Waters Canoe Area, Minnesota. *Quaternary Research New York*, **3**, 329–82.

Heinselman, M.L. (1981) Fire intensity and frequency as factors in the distribution and structure of northern ecosystems, in Proceedings of the Conference on Fire Regimes and Ecosystem Properties, (eds H.A. Mooney, T.M. Bonnicksen, N.L. Christensen, J.E. Lotan and W.A. Reiners), *USDA Forest Service, General Technical Report WO-26*, pp. 7–57.

Hibbert, A.R. (1979) Managing vegetation to increase flow in the Colorado River basin. *USDA Forest Service, General Technical Report RM-66.*

Hilbert, D.W. (1987) A model of life history strategies of chaparral shrubs in relation to fire frequency, in *Plant Responses to Stress*, (ed. J.D. Tenhunen), NATO ASI Series, vol. G15, pp. 597–606.

Hobbs, R.J. and Atkins, L. (1988) Spatial variability of experimental fires in south-west Western Australia. *Australian Journal of Ecology*, **13**, 295–9.

Hobbs, R.J. and Gimingham, C.H. (1984) Studies on fire in Scottish heathland communities II. Post-fire vegetation development. *Journal of Ecology*, **72**, 585–610.

Hobbs, R.J. and Gimingham, C.H. (1987) Vegetation, fire and herbivore interactions in heathland. *Advances in Ecological Research*, **16**, 87–173.

Hobbs, R.J., Mooney, H.A. (1985) Vegetative regrowth following cutting in the shrub *Baccharis pilularis* spp. *consanguinea* (DC) C.B. Wolf. *American Journal of Botany*, **72**, 514–19.

Hobbs, R.J., Currall, S.E.P. and Gimingham, C.H. (1984a) The use of 'thermocolor' pyrometers in the study of heath fire behaviour. *Journal of Ecology*, **72**, 241–50.

Hobbs, R.J., Mallik, A.U. and Gimingham, C.H. (1984b) Studies on fire in Scottish heathland communities III. Vital attributes of the species. *Journal of Ecology*, **72**, 963–76.

Hochberg, M.E., Menaut, J.C. and Gignoux, J. (1994) The influence of tree biology and fire in the spatial structure of the West African savannah. *Journal of Ecology*, **82**, 217–26.

Hodgkinson, K.C. (1986) Responses of rangeland plants to fire in water-limited environments, in *Rangelands: A Resource Under Siege*, (eds P.J. Joss, P.W. Lynch and O.B. Williams), Cambridge University Press, Cambridge, pp. 437–41.

Hodgkinson, K.C. (1991) Shrub recruitment response to intensity and season of fire in a semi-arid woodland. *Journal of Applied Ecology*, **28**, 60–70.

Hopkins, M.S., Ash, J., Graham, A.W. *et al.* (1993) Charcoal evidence of the spatial extent of the *Eucalyptus* woodland expansions and rainforest contractions in North Queensland during the late Pleistocene. *Journal of Biogeography*, **20**, 357–72.

Horn, H.S. (1971) *The Adaptive Geometry of Trees*. Princeton University Press, Princeton, New Jersey.

Horne, I.P. (1981) The frequency of veld fires in the Groot Swartberg Mountain Catchment Area, Cape Province. *South African Forestry Journal*, **118**, 56–60.

Horton, J.S. and Kraebel, C.J. (1955) Development of vegetation after fire in the chamise chaparral of southern California. *Ecology*, **36**, 244–62.

Hough, W.A. (1968) Carbohydrate reserves of saw-palmetto: seasonal variation and effects of burning. *Journal of Range Management*, **14**, 399–405.

Howe, H.F. (1994) Managing species diversity in Tallgrass Prairie: assumptions and implications. *Conservation Biology*, **8**, 691–704.

Hughes, F., Vitousek, P.M. and Tunison, T. (1991) Alien grass invasion and fire in the seasonal submontane zone of Hawai'i. *Ecology*, **72**, 743–7.

Hulbert, L.C. (1988) Causes of fire effects in tall grass prairie. *Ecology*, **69**, 46–58.

Humbert, H. (1927) La destruction d'une flore insulaire par le feu. Principaux aspects de la vegetation a Madagascar. *Memoires de l'Academie Malgache*, **5**, 1–78.

Huntly, N. (1991) Herbivores and the dynamics of communities and ecosystems. *Annual Review of Ecology and Systematics*, **22**, 477–503.

Huston, M. (1979) A general hypothesis of species diversity. *American Naturalist*, **113**, 81–101.

Jackson W.D. (1968) Fire, air, earth and water – an elemental ecology of Tasmania. *Proceedings of the Ecological Society of Australia*, **3**, 9–16.

James, S. (1984) Lignotubers and burls – their structure, function and ecological significance in Mediterranean ecosystems. *Botanical Review*, **50**, 225–66.

Jeffery, D.J., Holmes, P.M. and Rebelo, A.G. (1988) Effects of dry heat on seed germination in selected indigenous and alien legume species in South Africa. *South African Journal of Botany*, **54**, 28–34.

Johnson, E.A. (1978). Fire recurrence in the subarctic and its implications for vegetation composition. *Canadian Journal of Botany*, **57**, 1374–9.

Johnson, E.A. and Larsen, C.P.S. (1991) Climatically induced change in fire frequency in the southern Canadian Rockies. *Ecology*, **72**, 194–201.

Johnson, E.A. and van Wagner, E. (1984) The theory and use of two fire history models. *Canadian Journal of Forestry Research*, **15**, 214–20.

Jones, M.B. and Laude, H.M. (1960) Relationships between sprouting in chamise and the physiological condition of the plant. *Journal of Range Management*, **13**, 210–14.

Kauffman, J.B., Uhl, C. and Cummings, D.L. (1988) Fire in the Venezuelan Amazon 1: fuel biomass and fire chemistry in the evergreen rainforest of Venezuela. *Oikos*, **53**, 167–75.

Kayll, A.J. and Gimingham, C.H. (1965) Vegetative regeneration of *Calluna vulgaris* after fire. *Journal of Ecology*, **53**, 729–34.

Keane, R.F., Arno, F.A. and Brown, J.K. (1990). Simulating cumulative fire effects in Ponderosa pine/Douglas fir forests. *Ecology*, **71**, 189–203.

Keddy, P.A. (1989) *Competition*, Chapman & Hall, London.

Keddy, P.A. (1992) Assembly and response rules: two goals for predictive community ecology. *Journal of Vegetation Science*, **3**, 157–64.

Keeley, J.E. (1977a) Seed production, seed populations in soil, and seedling production after fire for two congeneric pairs of sprouting and non-sprouting chaparral shrubs. *Ecology*, **58**, 820–9.

Keeley, J.E. (1977b) Fire-dependent reproductive strategies in *Arctostaphylos* and *Ceanothus*, in Proceedings of the Symposium on Environmental Consequences of Fire and Fuel Manangement in Mediterranean Ecosystems, (eds H.A. Mooney and C.E. Conrad), *USDA Forest Service, General Technical Report WO-3*, pp. 391–6.

Keeley, J.E. (1986) Resilience of Mediterranean shrub communities to fires, in *Resilience in Mediterranean-type Ecosystems*, (eds B. Dell, A.J.M. Hopkins and B.B. Lamont), Junk, Dordrecht, pp. 95–112.

Keeley, J.E. (1991) Seed germination and life history syndromes in the California chaparral. *Botanical Review*, **57**, 81–116.

Keeley, J.E. (1993) Smoke-induced flowering in the fire lily *Cyrtanthus ventricosus. South African Journal of Botany*, **59**, 638.

Keeley, J.E. and Keeley, S.C. (1977) Energy allocation patterns of a sprouting and a non-sprouting species of *Arctostaphylos* in the California chaparral. *American Midland Naturalist*, **98**, 1–10.

Keeley, J.E. and Zedler, P.H. (1978) Reproduction of chaparral shrubs after fire: a comparison of sprouting and seeding strategies. *American Midland Naturalist*, **99**, 142–61.

Keeley, J.E., Morton, B.A., Pedrosa, A. and Trotter, P. (1985) Role of allelopathy, heat and charred wood in the germination of chaparral herbs and suffrutescents. *Journal of Ecology*, **73**, 445–58.

Keeley, S.C. and Pizzorno, M. (1986). Charred wood stimulated germination of two fire-following herbs of the California chaparral and the role of hemicellulose. *American Journal of Botany*, **73**, 1289–97.

Kelly, D. (1994) The evolutionary ecology of mast seeding. *Trends in Ecology and Evolution*, **9**, 465–70.

Kenkel, N.C. (1988) Pattern of self-thinning in jack pine: testing the random mortality hypothesis. *Ecology*, **69**, 1017–24.

Kenkel, N.C., Hoskins, J.A. and Hoskins, W.D. (1989) Local competition in a naturally established jack pine stand. *Canadian Journal of Botany*, **67**, 2630–5.

Kercher, J.R. and Axelrod, M.C. (1984) A process model of fire ecology and succession in a mixed-conifer forest. *Ecology*, **65**, 1725–42.

Kerr, L.R. (1925) The lignotubers of eucalypt seedlings. *Proceedings of the Royal Society of Victoria*, **37**, 79–97.

Kershaw, A.P. (1986) Climatic change and aboriginal burning in north-east Australia during the last two glacial/interglacial cycles. *Nature*, **322**, 47–9.

Kessel, S.R. (1990) An Australian geographical information system for natural area management. *International Journal of Geographical Information Systems*, **4**, 333–62.

Kessel, S.R., Potter, M.W., Bevins, C.D. *et al.* (1978) Analysis and application of forest fuels data. *Environmental Management*, **2**, 347–63.

Kessel, S.R., Good, R.B. and Potter, M.W. (1982) *Computer Modelling in Natural Area Management*, Special Publication 9, Australian National Parks and Wildlife Service.

Kessel, S.R., Good, R.B. and Hopkins, A.J.M. (1984) Implementation of two new resource management information systems in Australia. *Environmental Management*, **8**, 251–70.

Koechlin, J., Guillaumet, J.L. and Morat, P. (1974) *Flore et vegetation de Madagascar*, Vaduz, Cramer.

Kruger, F.J. (1977) Ecology of Cape fynbos in relation to fire, in Proceedings of the Symposium on Environmental Consequences of Fire and Fuel Management in Mediterranean Ecosystems, (eds H.A. Mooney and C.E. Conrad), *USDA Forest Service, General Technical Report WO-3*, pp. 230–44.

Kruger, F.J. (1983) Plant community diversity and dynamics in relation to fire, in *Mediterranean-Type Ecosystems: The Role of Nutrients*, (eds F.J. Kruger, D. Mitchell and J. Jarvis), Springer Verlag, Berlin, pp. 446–72.

Kruger, F.J. (1986) Succession after fire in selected fynbos communities of the South-Western Cape. Unpublished Ph.D. thesis, University of the Witwatersrand, South Africa.

Kruger, F.J. and Bigalke, R.C. (1984) Fire in fynbos, in *Ecological Effects of Fire in South African Ecosystems*, (eds P. de V. Booysen and N.M. Tainton), Springer-Verlag, Berlin, pp. 67–114.

Lamont, B.B. (1985) The comparative reproductive biology of three *Leucospermum* species (Proteaceae) in relation to fire responses and breeding system. *Australian Journal of Botany*, **33**, 139–45.

Lamont, B.B. and Bergl, S.M. (1991) Water relations, shoot and root architecture, and phenology of three co-occurring *Banksia* species: no evidence for niche differentiation in the pattern of water use. *Oikos*, **60**, 291–8.

Lamont, B.B. and Cowling, R.M. (1984) Flammable infructescences in *Banksia*: a fruit-opening mechanism. *Australian Journal of Ecology*, **9**, 295–6.

Lamont, B.B., Collins, B.G. and Cowling, R.M. (1985) Reproductive biology of the Proteaceae in Australia and South Africa. *Proceedings of the Ecological Society of Australia*, **14**, 213–24.

Lamont, B.B., Enright, N.J. and Bergl, S.M. (1989) Coexistence and competitive exclusion of *Banksia hookeriana* in the presence of congeneric seedlings along a topographic gradient. *Oikos*, **56**, 39–42.

Lamont, B.B., Connell, S.W. and Bergl, S.M. (1991a) Seed bank and population dynamics of *Banksia cuneata*: the role of time, fire and moisture. *Botanical Gazette*, **152**, 114–22.

Lamont, B.B., Le Maitre, D.C., Cowling, R.M. and Enright, N.J. (1991b) Canopy seed storage in woody plants. *Botanical Review*, **57**, 277–317.

Lamont B.B., Witkowski, E.T.F. and Enright, N.J. (1993) Post-fire litter microsites: safe for seeds, unsafe for seedlings. *Ecology*, **74**, 501–12.

Landhausser, S.M. and Wein, R.W. (1993) Postfire vegetation recovery and tree establishment at the Arctic treeline: climate-change-vegetation-response hypotheses. *Journal of Ecology*, **81**, 665–72.

Laude, H.M., Jones, M.B. and Moon, W.E. (1961) Annual variability in indicators of sprouting potential in chamise. *Journal of Range Management*, **14**, 323–6.

Leishman, M.R. and Westoby, M. (1992) Classifying plants into groups on the basis of associations of individual traits – evidence from Australian semi-arid woodlands. *Journal of Ecology*, **80**, 417–24.

Le Maitre, D.C. (1984) A short note on seed predation in *Watsonia pyramidata* (Andr.) Stapf in relation to season of burn. *Journal of South African Botany*, **50**, 407–15.

Le Maitre, D.C. (1988) Effects of season of burn on the regeneration of two Proteaceae with soil-stored seed. *South African Journal of Botany*, **54**, 575–80.

Le Maitre, D.C. (1992) The relative advantages of seeding and sprouting in fire-prone environments: a comparison of life histories of *Protea neriifolia* and *Protea nitida*, in *Fire in South African Mountain Fynbos*, (eds B.W. van Wilgen, D.M. Richardson, F.J. Kruger and H.J. van Hensbergen), Ecological Studies 93, Springer-Verlag, Berlin, pp. 123–43.

Le Maitre, D.C. and Brown, P.J. (1992) Life cycles and fire-stimulated flowering in geophytes, in *Fire in South African Mountain Fynbos*, (eds B.W. van Wilgen, D.M. Richardson, F.J. Kruger and H.J. van Hensbergen), Ecological Studies 93, Springer Verlag, Berlin, pp. 145–60.

Le Maitre, D.C., Midgley, J.J. (1992) Plant reproductive ecology, in *The Ecology of Fynbos: Nutrients, Fire and Diversity*, (ed. R.M. Cowling), Oxford University Press, Cape Town, pp. 135–74.

Le Maitre, D.C., Jones, C.A. and Forsyth, G.G. (1992) Survival of eight woody sprouting species following an autumn fire in Swartboskloof, Cape Province, South Africa. *South African Journal of Botany*, **58**, 405–13.

Le Roux, C.J.G. (1989) Die invloed van brand en maai op onbeweide langrasveld in Natal. (The effect of burning and mowing on ungrazed tall grassveld of Natal.) *Journal of the Grasslands Society of South Africa*, **6**, 59–64.

Levine, J.S. (1991a). Introduction, in *Global Biomass Burning: Atmospheric, Climatic and Biospheric Implications*, (ed. J.S. Levine), MIT Press, Cambridge, MA, pp. xxv–xxx.

Levine, J.S. (1991b) *Global Biomass Burning: Atmospheric, Climatic and Biospheric Implications*, MIT Press, Cambridge, MA.

Linder, H.P. (1995) *Ceratocaryum pulchrum*, a new restioid from the Bredasdorp plains. *South African Journal of Botany* (in press).

Lloret, F. and Lopez-Soria, L. (1993) Resprouting of *Erica multiflora* after experimental fire treatments. *Journal of Vegetation Science*, **4**, 367–74.

Loehle, C. (1988) Tree life history strategies: the role of defenses. *Canadian Journal of Forestry*, **18**, 209–22.

Lomnicki, A. (1988) *Population Ecology of Individuals*, Princeton University Press, Princeton, New Jersey.

Ludwig, J.A. (1990) SHRUBKILL: a decision support system for management burns in Australian savannas. *Journal of Biogeography*, **17**, 547–50.

Luger, A.D. and Moll, E.J. (1992) Fire protection and Afromontane forest expansion in Cape fynbos. *Biological Conservation*, **64**, 51–6.

Luke, R.H. and McArthur, A.G. (1978) *Bushfires in Australia*, Australian Government Publishing Service, Canberra.

Luken, J.O. (1990) *Directing Ecological Succession*, Chapman & Hall, London.

McArthur, A.G. (1966) Weather and grassland fire behaviour. *Commonwealth of Australia, Forestry and Timber Bureau Leaflet, 100.*

McArthur, A.G. (1967) Fire behaviour in eucalypt forest. *Commonwealth of Australia, Forestry and Timber Bureau Leaflet, 107.*

MacLean, D.A., Woodley, S.J., Weber, M.G. and Wein, R.W. (1983) Fire and nutrient cycling, in *The Role of Fire in Northern Circumpolar Ecosystems*, (eds R.W. Wein and D.A. MacLean), SCOPE, Wiley & Sons, New York, pp. 11–132.

McMaster G.S. and Zedler, P.H. (1981) Delayed seed dispersal in *Pinus torreyana* (Torrey Pine). *Oecologia*, **51**, 62–6.

McNaughton, S.J. (1984) Grazing lawns: animals in herds, plant form, and coevolution. *American Naturalist*, **124**, 863–86.

McNaughton, S.J., Ruess, R.W. and Seagle, S.W. (1988) Large mammals and process dynamics in African ecosystems. *Bioscience*, **38**, 794–800.

McPherson, J.K. and Muller, C.H. (1969) Allelopathic effects of *Adenostoma fasciculatum*, 'chamise', in the Californian chaparral. *Ecological Monographs*, **39**, 177–98.

Main, A.R. (1986) Resilience at the level of the individual animal, in *Resilience in Mediterranean-type Ecosystems*, (eds B. Dell, A.J.M. Hopkins and B.B. Lamont), Junk, Dordrecht, pp. 83–94.

Malanson, G.P. and Trabaud, L. (1988) Vigour of post-fire resprouting by *Quercus coccifera* L. *Journal of Ecology*, **76**, 351–65.

Mallik, A.U. and Gimingham, C.H. (1985) Ecological effects of heather burning II. Effects on seed germination and vegetative regeneration. *Journal of Ecology*, **73**, 633–44.

Mallik, A.U., Hobbs, R.J. and Legg, C.J. (1984) Seed dynamics in *Calluna–Arctostaphylos* heath in north-eastern Scotland. *Journal of Ecology*, **72**, 855–71.

Maltby, E., Legg, C.J. and Proctor, M.C.F. (1990) The ecology of severe moorland fire on the North York Moors: effects of the 1976 fires, and subsequent surface and vegetation development. *Journal of Ecology*, **78**, 490–518.

Manders, P.T. (1987) A transition matrix model of the population dynamics of the Clanwilliam cedar (*Widdringtonia cedarbergensis*) in natural stands subject to fire. *Forest Ecology and Management*, **20**, 171–86.

Manders, P.T. (1990) Fire and other variables as determinants of forest/fynbos boundaries in the Cape Province. *Journal of Vegetation Science*, **1**, 483–90.

Manders, P.T., Richardson, D.M. and Masson, P.H. (1992) Is fynbos a stage in succession to forest? Analysis of the perceived ecological distinction between two communities, in *Fire in South African Mountain Fynbos*, (eds B.W. van Wilgen, D.M. Richardson, F.J. Kruger and H.J. van Hensbergen), Ecological Studies 93, Springer-Verlag, Berlin, pp. 81–107.

Manly, B.F.J. (1990) *Stage-structured populations*, Chapman & Hall, London.

Mark, A.F. (1965) Flowering, seeding and seedling establishment of narrow-leaved snow tussock, *Chionochloa rigida. New Zealand Journal of Botany*, **3**, 180–93.

May, R.M. (1974) Biological populations with nonoverlapping generations: stable points, stable cycles, and chaos. *Science*, **186**, 645–7.

Maze, K.E. and Bond, W.J. (1995). Are protea populations seed limited? Implications for wild flower harvesting in Cape fynbos. *Australian Journal of Ecology*, (in press).

Menaut, J.C., Gignoux, J., Prado, C. and Clobert, J. (1990) Tree community dynamics in a humid savanna of the Cote-d'Ivoire: modelling the effects of fire and competition with grass and neighbours. *Journal of Biogeography*, **17**, 471–81.

Meyer, G.A., Wells, S.G., Balling, R.C. and Jull, A.J.T. (1992) Response of alluvial systems to fire and climate change in Yellowstone National Park. *Nature*, **357**, 147–50.

Midgley, J.J. (1987) The derivation, utility and implications of a divergence index for the genus *Leucadendron* (Proteaceae). *Botanical Journal of the Linnaean Society*, **95**, 137–52.

Midgley, J.J. (1988) Mortality of Cape Proteaceae seedlings during their first summer. *South African Forestry Journal*, **145**, 9–12.

Midgley, J.J. (1989) Season of burn of serotinous fynbos Proteaceae: a critical review and further data. *South African Journal of Botany*, **55**, 165–70.

Midgley, J.J. and Clayton, P. (1990) Short-term effects of an autumn fire on small mammal populations in southern Cape coastal mountain fynbos. *South African Forestry Journal*, **153**, 27–30.

Midgley, J.J., Hoekstra, T. and Bartholomew, R. (1989) The effect of date of planting on field establishment of serotinous Cape Proteaceae. *Vegetatio*, **79**, 185–92.

Mills, J.N. (1983) Herbivory and seedling establishment in post-fire southern California chaparral. *Oecologia*, **60**, 267–70.

Mills, J.N. (1986) Herbivores and early post-fire succession in southern Californian chaparral. *Ecology*, **67**, 1637–49.

Mitchell, D.T. and Coley, P.G.F. (1987) Litter production and decomposition from shrubs of *Protea repens* growing in sand plain lowland and mountain fynbos, southwestern Cape. *South African Journal of Botany*, **53**, 25–31.

Moll, E.J., McKenzie, B. and McLachlan, D. (1980) A possible explanation for the lack of trees in the fynbos, Cape Province, South Africa. *Biological Conservation*, **17**, 221–8.

Molofsky, J. and Augspurger, C.K. (1992) The effect of leaf litter on early seedling establishment in a tropical forest. *Ecology*, **73**, 68–77.

Mooney, H.A. (ed.) (1977) *Convergent Evolution in Chile and California: Mediterranean Climate Ecosystems*, Series 5, United States/IBP Synthesis Dowden, Hutchinson and Ross, Stroudsburg, PA.

Mooney, H.A., Kummerow, J., Johnson, A.W. *et al.* (1977) The producers – their resources and adaptive responses, in *Convergent Evolution in Chile and California: Mediterranean Climate Ecosystems*, (ed. H.A. Mooney), Series 5, United States/ IBP Synthesis, Dowden, Hutchinson and Ross, Stroudsburg, PA, pp. 85–143.

Moore, A.D. and Noble, I.R. (1990) An individualistic model of vegetation stand dynamics. *Journal of Environmental Management*, **31**, 61–81.

Moore, C.T. (1972) Man and fire in the central North American grassland 1535–1890: A documentary historical geography. Ph.D. thesis, University of California, Los Angeles.

Moreno, J.M. and Oechel, W.C. (1991a) Fire intensity effects on germination of shrubs and herbs in southern California chaparral. *Ecology*, **72**, 1993–2004.

Moreno, J.M. and Oechel, W.C. (1991b) Fire intensity and herbivory effects on postfire resprouting of *Adenostoma fasciculatum* in southern California chaparral. *Oecologia*, **85**, 429–33.

Moreno, J. M. and Oechel, W.C. (1993) Demography of *Adenostoma fasciculatum* after fires of different intensities in southern California chaparral. *Oecologia*, **96**, 95–101.

Mount, A.B. (1964) The interdependence of the eucalypts and forest fires in southern Australia. *Australian Forestry*, **30**, 231–41.

Muir, P.S. and Lotan, J.E. (1985) Disturbance history and serotiny of *Pinus contorta* in western Montana. *Ecology*, **66**, 1658–68.

Munoz, M.R. and Fuentes, E.R. (1989) Does fire induce shrub germination in the Chilean matorral? *Oikos*, **56**, 177–81.

Mutch, R.W. (1970) Wildland fires and ecosystems – a hypothesis. *Ecology*, **51**, 1046–51.

Naveh, Z. (1975) The evolutionary significance of fire in the Mediterranean region. *Vegetatio*, **29**, 199–208.

Ne'eman, G., Lahav, H. and Izhaki, I. (1992) Spatial pattern of seedlings 1 year after fire in a Mediterranean pine forest. *Oecologia*, **91**, 365–70.

Newman, E.I. (1993) *Applied Ecology*, Blackwell, Oxford.

Noble, I.R. (1984) Mortality of lignotuberous seedlings of *Eucalyptus* species after an intense fire in montane forest. *Australian Journal of Ecology*, **9**, 47–50.

Noble, I.R. (1987) The role of expert systems in vegetation science. *Vegetatio*, **69**, 115–21.

Noble, I.R. and Slatyer, R.O. (1980) The use of vital attributes to predict successional changes in plant communities subject to recurrent disturbances. *Vegetatio*, **43**, 5–21.

Noble, I.R., Bary, G.A.V. and Gill, A.M. (1980) McArthur's fire-danger meters expressed as equations. *Australian Journal of Ecology*, **5**, 201–3.

Norton, D.A. and Kelly, D. (1988) Mast seeding over 33 years by *Dacrydium cupressinum* Lamb. (rimu) (Podocarpaceae) in New Zealand: the importance of economies of scale. *Functional Ecology*, **2**, 399–408.

Norton Griffiths, M. (1979) The influence of grazing, browsing and fire on the vegetation dynamics of the Serengeti, in *Serengeti: Dynamics of an Ecosystem*, (eds A.R.E. Sinclair and M. Norton Griffiths), University of Chicago Press, Chicago, pp. 310–52.

O'Dowd, D.J. and Gill, A.M. (1984) Predator satiation and site alteration following fires: mass reproduction of alpine ash (*Eucalyptus delegatensis*) in south-eastern Australia. *Ecology*, **65**, 1052–66.

Oksanen, L. (1988) Ecosystem organization: mutualism and cybernetics or plain Darwinian struggle for existence? *American Naturalist*, **131**, 424–44.

Oksanen, L., Fretwell, S.D., Arruda, J. and Niemela, P. (1981) Exploitation ecosystems in gradients of primary productivity. *American Naturalist*, **118**, 240–61.

Old, S. (1969) Microclimate, fire and plant production in an Illinois prairie. *Ecological Monographs*, **39**, 355–84.

Pacala, S.W. and Silander, J.A. (1985) Neighbourhood models of plant population dynamics. I, Single species models of annuals. *American Naturalist*, **125**, 385–411.

Papavassiliou, S. and Arianoutsou, M. (1993) Regeneration of the leguminous herbaceous vegetation following fire in a *Pinus halepensis* forest of Attica, Greece, in *Fire in Mediterranean Ecosystems*, (eds L. Trabaud and R. Prodon), Ecosystems Research Report, Commission of the European Communities, pp. 119–25.

Papio, C. and Trabaud, L. (1991) Comparative study of the aerial structure of five shrubs of Mediterranean shrublands. *Forest Science*, **37**, 146–59.

Parker, V.T. and Kelly, V.R. (1989) Seed banks in California chaparral and other Mediterranean climate shrublands, in *Ecology of Soil Seed Banks*, (eds M.A. Leck, V.T. Parker and R.L. Simpson), Academic Press, San Diego, pp. 231–56.

Pate, J.S., Froend, R.H., Bowen, B.J. *et al.* (1990) Seedling growth and storage characteristics of seeder and resprouter species of Mediterranean-type ecosystems of S.W. Australia. *Annals of Botany*, **65**, 585–601.

Pate, J.S., Meney, K.A. and Dixon, K.W. (1991) Contrasting growth and morphological characteristics of fire-sensitive (obligate seeder) and fire-resistant (resprouter) species of Restionaceae (S. hemisphere restiads) from south-western Australia. *Australian Journal of Botany*, **39**, 505–25.

Peet, R.K. (1992) Community structure and ecosystem function, in *Plant Succession: Theory and Prediction*, (eds D.C. Glenn-Lewin, R.K. Peet and T.T. Veblen),

Chapman & Hall, London, pp. 103–51.
Pellew, R.A.P. (1983) The impact of elephant, giraffe and fire upon the *Acacia tortilis* woodlands of the Serengeti. *African Journal of Ecology*, **21**, 41–74.
Perry, D.A. and Lotan, J.E. (1979) A model for fire selection for serotiny in lodgepole pine. *Evolution*, **33**, 958–68.
Peterson, D.L. and Ryan, K.C. (1986) Modelling postfire conifer mortality for long range planning. *Environmental Management*, **10**, 797–808.
Phillips, J.F.V. (1934–35) Succession, development, the climax and the complex organism: an analysis of concepts. Pt I, *Journal of Ecology* **22**, 88–105; Pt II, **23**, 210–23; Pt III, **23**, 554–71.
Philpot, C.W. (1969) Seasonal changes in heat content and ether extractive content of chamise. *USDA Forest Service Research Paper INT-61.*
Philpot, C.W. (1977) Vegetative features as determinants of fire frequency and intensity, in Proceedings of the Symposium on the Environmental Consequences of Fire and Fuel Management in Mediterranean Ecosystems, (eds H.A. Mooney and C.E. Conrad), *USDA Forest Service, Technical Report WO-3*, pp. 12–16.
Pickett, S.T.A. and White, P.S. (eds) (1985) *The Ecology of Natural Disturbance*, Academic Press, New York.
Pickett, S.T.A., Collins, S.L. and Armesto, J.J. (1987) Models, mechanisms and pathways of succession. *Botanical Review*, **53**, 335–71.
Pickett, S.T.A., Parker, V.T. and Fiedler, P.L. (1992) The new paradigm in ecology: implications for conservation biology above the species level, in *Conservation Biology. The Theory and Practice of Nature Preservation and Management*, (eds P.L. Fiedler and S.K. Jain), Chapman & Hall, London, pp. 65–88.
Pierce, S.M. and Cowling, R.M. (1991) Dynamics of soil-stored seed banks of six shrubs in fire-prone dune fynbos. *Journal of Ecology*, **79**, 731–47.
Poissonet, J., Poissonet, P. and Thiault, M. (1981) Development of flora, vegetation and grazing value in experimental plots of a *Quercus coccifera* garrigue. *Vegetatio*, **46**, 93–104.
Portlock, C.C., Shea, S.R., Majer, J.D. and Bell, D.T. (1990) Stimulation of germination of *Acacia pulchella*: laboratory basis for forest management options. *Journal of Applied Ecology*, **27**, 319–24.
Prentice, I.C., Monserud, R.A., Smith, T.M. and Emanuel, W.R. (1993) Modelling large-scale vegetation dynamics, in *Vegetation Dynamics and Global Change*, (eds A.M. Solomon and H.H. Shugart), Chapman & Hall, London, pp. 235–50.
Preston-Mafham, K. (1991) *Madagascar: A Natural History*, Struik, Cape Town.
Putman, R.J. and Wratten, S.D. (1984) *Principles of Ecology*, University of California Press, Berkeley.
Pyke, G.H. (1983) Relationship between time since the last fire and flowering in *Telopea speciosissima* and *Lambertia formosa*. *Australian Journal of Botany*, **31**, 293–6.
Pyne, S.J. (1982) *Fire in America: A Cultural History of Wildland and Rural Fire*, Princeton University Press, Princeton, New Jersey.
Pyne, S.J. (1984) *Introduction to Wildland Fire – Fire Management in the United States*, Wiley, New York.
Pyne, S.J. (1989) The summer we let wild fire loose: How 'natural' were the fires of 1988? *Natural History*, **8**, 45–9.

Pyne, S.J. (1990) Fire conservancy: The origins of wildland fire protection in British India, America and Australia, in *Fire in the Tropical Biota*, (ed. J.G. Goldammer), Ecological Studies 84, Springer-Verlag, Berlin, pp. 319–36.
Pyne, S.J. (1991) *Burning Bush: A Fire History of Australia*, Henry Holt, New York.
Quinlivan, B.J. (1971) Seed coat impermeability in legumes. *Journal of the Australian Institute of Agricultural Science*, **37**, 283–95.
Radley, J. (1965) Significance of major moorland fires. *Nature*, **205**, 1254–9.
Rebertus, A.J., Williamson,G.B. and Moser, E.B. (1989) Longleaf pine pyrogenicity and turkey oak mortality in Florida xeric sandhills. *Ecology*, **70**, 60–70.
Rees, M. and Crawley, M.J. (1989) Growth, reproduction and population dynamics. *Functional Ecology*, **3**, 645–53.
Reich, P.B., Uhl, C., Walters, M.B. and Ellsworth, D.S. (1991) Leaf life span as a determinant of leaf structure and function among 23 Amazonian tree species. *Oecologia*, **86**, 16–24.
Rice, S.K. (1993) Vegetation establishment in post-fire *Adenostoma* chaparral in relation to fine-scale pattern in fire intensity and soil nutrients. *Journal of Vegetation Science*, **4**, 115–24.
Richards, J.H. (1986) Plant response to grazing: the role of photosynthetic capacity and stored carbon reserves, in *Rangelands: A Resource Under Siege*, (eds P.J. Joss, P.W. Lynch and O.B. Williams), Cambridge University Press, Cambridge, pp. 428–30.
Richardson, D.M., Macdonald, I.A.W., Holmes, P.M. and Cowling, R.M. (1992) Plant and animal invasions, in *The Ecology of Fynbos: Nutrients, Fire and Diversity*, (ed. R.M. Cowling), Oxford University Press, Cape Town, pp. 271–308.
Richardson, D.M., van Wilgen, B.W., Le Maitre, D.C. *et al.* (1994) A computer-based system of fire management in the mountains of the Cape Province, South Africa. *International Journal of Wildland Fire*, **4**, 17–32.
Ricklefs, R.E. (1990) *Ecology*, 3rd edn, Freeman, New York.
Riggan, P.J., Goode, S., Jacks, P.M. and Lockwood, R.N. (1988) Interaction of fire and community development in chaparral of southern California. *Ecological Monographs*, **58**, 155–76.
Romme, W.H. (1982) Fire and landscape diversity in subalpine forests of Yellowstone National Park. *Ecological Monographs*, **52**, 199–221.
Rose-Innes, R. (1972) Fire in West African vegetation. *Proceedings of the Tall Timbers Fire Ecology Conference*, **11**, 147–73.
Rothermel, R.C. (1972) A mathematical model for predicting fire spread in wildland fuels. *USDA Forest Service Research Paper INT-115.*
Rothermel, R.C. (1991) Predicting behaviour of the 1988 Yellowstone fires: projections versus reality. *International Journal of Wildland Fire*, **1**, 1–10.
Rothermel, R.C. and Philpot, C.W. (1973) Predicting changes in chaparral flammability. *Journal of Forestry*, **71**, 640–3.
Rowe, J.S. (1983) Concepts of fire effects on plant individuals and species, in *The Role of Fire in Northern Circumpolar Ecosystems*, (eds R.W. Wein and D.A. MacLean), SCOPE 18, Wiley, New York, pp. 135–54.
Rundel, P.W. (1981) Structural and chemical components of flammability, in Proceedings of the Conference on Fire Regimes and Ecosystem Properties, (eds H.A. Mooney, T.M. Bonnicksen, N.L. Christensen, J.E. Lotan and W.A. Reiners), *USDA Forest Service General Technical Report WO-86*, pp. 183–207.
Rundel, P.W., Baker, G.A., Parsons, D.J. and Stohlgren, T.J. (1987) Postfire demog-

raphy of resprouting and seedling establishment by *Adenostoma fasciculatum* in the California chaparral, in *Plant Response to Stress – Functional Analysis in Mediterranean Ecosystems*, (eds J.D. Tenhunen, F.M. Catarino, O.L. Lange and W.C. Oechel), Springer-Verlag, Berlin, pp. 575–96.

Ryan, K.C. and Reinhardt, E.D. (1988) Predicting postfire mortality of seven western conifers. *Canadian Journal of Forest Research*, **18**, 1291–7.

San Jose, J.J. and Farias, M.R. (1983) Changes in tree density and species composition in a protected *Trachypogon* savanna, Venezuela. *Ecology*, **64**, 447–53.

Sarmiento, G. (1983) The savannas of tropical America, in *Ecosystems of the World 13, Tropical Savannas*, (ed. F. Bourliere), Elsevier, Amsterdam, pp. 245–88.

Sauer, C.O. (1950) Grassland climax, fire and man. *Journal of Range Management*, **8**, 117–21.

Schlesinger, W.H. and Gill, D.S. (1978) Demographic studies of the chaparral shrub, *Ceanothus megacarpus*, in the Santa Ynez mountains, California. *Ecology*, **59**, 1256–63.

Schlesinger, W.H. and Hasey, M.M. (1981) Decomposition of chaparral shrub foliage: losses of organic and inorganic constituents from deciduous and evergreen leaves. *Ecology*, **62**, 762–74.

Schullery, P. (1989) The fires and fire policy. *Bioscience*, **39**, 686–94.

Scott, D.F. and van Wyk, D.B. (1990) The effects of wildfire on soil wettability and hydrological behaviour of an afforested catchment. *Journal of Hydrology*, **121**, 239–56.

Scott, D.F. and van Wyk, D.B. (1992) The effects of fire on soil water repellency, catchment sediment and yields and streamflow, in *Fire in South African Mountain Fynbos*, (eds B.W. van Wilgen, D.M. Richardson, F.J. Kruger and H.J. van Hensbergen), Ecological Studies 93, Springer-Verlag, Berlin, pp. 216–39.

Shea, S.R., McCormick, J. and Portlock, C.C. (1979) The effects of fires on regeneration of leguminous species in the northern jarrah (*Eucalyptus marginata* Sm.) forest of Western Australia. *Australian Journal of Ecology*, **4**, 195–205.

Shipley, B., Keddy, P.A., Moore, D.R.J. and Lemkt, K. (1989) Regeneration and establishment strategies of emergent macrophytes. *Journal of Ecology*, **77**, 1093–110.

Shugart, H.H. (1984) *A Theory of Forest Dynamics*, Springer Verlag, New York.

Shugart, H.H. and Noble, I.R. (1981) A computer model of succession and fire response of the high altitude *Eucalyptus* forest of the Brindabella Range, Australian Capital Territory. *Australian Journal of Ecology*, **6**, 149–64.

Silva, J.F., Raventos, J., Caswell, H. and Trevisan, M.C. (1991) Population responses to fire in a tropical savanna grass, *Andropogon semiberbis*: a matrix model approach. *Journal of Ecology*, **79**, 345–56.

Silvertown, J. (1980) The evolutionary ecology of mast seeding in trees. *Biological Journal of the Linnaean Society*, **14**, 235–50.

Silvertown, J. (1991) Modularity, reproductive thresholds and plant population dynamics. *Functional Ecology*, **5**, 577–82.

Singh, G., Kershaw, A.P. and Clark, R. (1981) Quaternary vegetation and fire history in Australia, in *Fire and the Australian Biota*, (eds A.M. Gill, R.H. Groves and I.R. Noble), Australian Academy of Science, Canberra, pp. 23–54.

Smith, R.E., van Wilgen, B.W., Forsyth, G.G. and Richardson, D.M. (1992) Coexistence of seeders and sprouters in a fire-prone environment: the role of ecophysiology and soil moisture, in *Fire in South African Mountain Fynbos*, (eds

B.W. van Wilgen, D.M. Richardson, F.J. Kruger and H.J. van Hensbergen), Ecological Studies 93, Springer-Verlag, Berlin, pp. 108–22.

Smith, R.L. (1990) *Ecology and Field Biology*, 4th edn, Harper and Row, New York.

Smith, T.M. and Goodman, P.S. (1987) Successional dynamics in an *Acacia nilotica – Euclea divinorum* savannah in southern Africa. *Journal of Ecology*, **75**, 603–10.

Smith, T.M., Shugart, H.H., Woodward, F.I. and Burton, P.J. (1993) Plant functional types, in *Vegetation Dynamics and Global Change*, (eds A.M. Solomon and H.H. Shugart), Chapman & Hall, London, pp. 272–92.

Snyder, J.R. (1984) The role of fire: Mutch ado about nothing? *Oikos*, **43**, 404–5.

Sork, V.L. (1983) Mammalian seed dispersal of pignut hickory during three fruiting seasons. *Ecology*, **64**, 1049–56.

Sousa, W.P. (1984) The role of disturbance in natural communities. *Annual Review of Ecology and Systematics*, **15**, 353–91.

Specht, R.L. (1981) Responses to fires in heathlands and related shrublands, in *Fire and the Australian Biota*, (eds A.M. Gill, R.H. Groves and I.R. Noble), Australian Academy of Science, Canberra, pp. 395–415.

Specht, R.L. (ed.) (1988) *Mediterranean-type Ecosystems: A Data Source Book*, Kluwer, Dordrecht.

Specht, R.L. and Moll, E.J. (1983) Heathlands and sclerophyllous shrublands – an overview, in *Mediterranean-Type Ecosystems: the Role of Nutrients*, (eds F.J. Kruger, D. Mitchell and J. Jarvis), Springer Verlag, Berlin, pp. 41–65.

Specht, R.L., Rayson, P. and Jackman, M.E. (1958) Dark island heath (Ninety mile plain, SOUTH). VI. Pyric succession changes in composition, coverage, dry weight and mineral nutrient status. *Australian Journal of Botany*, **6**, 59–88.

Sprugel, D.G. (1991) Disturbance, equilibrium, and environmental variability: What is 'natural' vegetation in a changing environment? *Biological Conservation*, **58**, 1–18.

Spurr, S.H. (1954) The forests of Itassa in the nineteenth century as related to fire. *Ecology*, **35**, 21–5.

Starfield, A.M. and Bleloch, A.L. (1983) Expert systems: An approach to problems in ecological management that are difficult to quantify. *Journal of Environmental Management*, **16**, 261–8.

Starfield, A.M., Cumming, D.H.M., Taylor, R.D. and Quadling, M.S. (1993) A frame-based paradigm for dynamic ecosystem models. *AI Applications*, **7**, 1–13.

Stearns, S.C. (1977) The evolution of life history traits. *Annual Review of Ecology and Systematics*, **8**, 145–71.

Stebbins, G.L. (1981) Coevolution of grasses and herbivores. *Annals of the Missouri Botanical Gardens*, **68**, 75–86.

Stewart, O.C. (1956) Fire as the first great force employed by man, in *Man's Role in Changing the Face of the Earth*, (ed. W.L. Thomas), University of Chicago Press, Chicago, pp. 115–33.

Stiling, P.D. (1992) *Introductory Ecology*, Prentice Hall, Englewood Cliffs, New Jersey.

Stock, W.D. and Lewis, O.A.M. (1986) Soil nitrogen and the role of fire as a mineralizing agent in a South African coastal fynbos ecosystem. *Journal of Ecology*, **74**, 317–28.

Stoddart, L.A., Smith, A.D. and Box, T.W. (1975) *Range Management*, 3rd edn, McGraw-Hill, New York.

Stohlgren, T.J. and Rundel, P.W. (1986) A population model for a long-lived, resprouting chaparral shrub: *Adenostoma fasciculatum*. *Ecological Modelling*, **34**, 245–57.

Stott, P.A. (1988) The forest as Phoenix: towards a biogeography of fire in mainland South East Asia. *Geographical Journal*, **154**, 337–50.

Stott, P.A., Goldammer, J.G. and Werner, W.L. (1990) The role of fire in the tropical lowland deciduous forests of Asia, in *Fire in the Tropical Biota*, (ed. J.G. Goldammer), Springer Verlag, Berlin, pp. 32–44.

Svejcar, T.J. (1990) Response of *Andropogon gerardii* to fire in the Tallgrass prairie, in *Fire in North American Tallgrass Prairies*, (eds S.L. Collins and L.L. Wallace), University of Oklahoma Press, Norman, pp. 19–27.

Swetnam, T.W. and Betancourt, J.L. (1990) Fire-southern oscillation relations in the southwestern United States. *Science*, **249**, 961–1076.

Symonides, E., Silvertown, J. and Andreasen, V. (1986) Population cycles caused by over-compensating density dependence in an annual plant. *Oecologia*, **71**, 156–8.

Tainton, N.M. (ed.) (1981) *Veld and Pasture Management in South Africa*, Shuter and Shooter, Pietermaritzburg, Natal.

Tainton, N.M. and Mentis, M.T. (1984) Fire in grassland, in *Ecological Effects of Fire in South African Ecosystems*, (eds P. de V. Booysen and N.M. Tainton), Springer Verlag, Berlin, pp. 115–48.

Teich, A.H. (1970) Cone serotiny and inbreeding in natural populations of *Pinus banksiana* and *Pinus contorta*. *Canadian Journal of Botany*, **48**, 1805–9.

Thanos, C.A., Georghiou, K., Kadis, C. and Pantazi, C. (1994) Cistaceae: a family with hard seeds. *Israeli Journal of Botany*, **41**, 151–263.

Thomas, C.M. and Davis, S.D. (1989) Recovery patterns of three chaparral shrub species after wildfire. *Oecologia*, **80**, 309–20.

Thrall, P.H., Pacala, S.W. and Silander, J.A. (1989) Oscillatory dynamics in populations of an annual weed species *Abutilon theophrasti*. *Journal of Ecology*, **77**, 1135–49.

Tilman, D. (1982) *Resource Competition and Community Structure*, Princeton University Press, Princeton, New Jersey.

Tilman, D. (1988) *Plant Strategies and the Dynamics and Structure of Plant Communities*, Princeton University Press, Princeton, New Jersey.

Tilman, D. (1990) Constraints and trade-offs: toward a predictive theory of competition and succession. *Oikos*, **58**, 3–15.

Tilman, D. (1994) Competition and biodiversity in spatially structured habitats. *Ecology*, **75**, 2–16.

Tilman, D. and Wedin, D. (1991) Oscillations and chaos in the dynamics of a perennial grass. *Nature*, **353**, 653–5.

Torn, M.S. and Fried, J.S. (1992) Predicting the impacts of global warming on wildland fire. *Climatic Change*, **21**, 257–74.

Towne, G. and Owensby, C. (1984) Long term effects of annual burning at different dates on ungrazed Kansas tallgrass prairie. *Journal of Range Management*, **37**, 92–7.

Trabaud, L. (1984) Fire adaptation strategies of plants in the French Mediterranean region, in *Being Alive on Land*, (eds N.S. Margaris, M. Arianoutsou-Faraggitaki and W.C. Oechel), Tasks for Vegetation Science 13, Kluwer, Dordrecht, pp. 63–9.

Trabaud, L. (1991) Fire regimes and phytomass growth dynamics in a *Quercus coccifera* garrigue. *Journal of Vegetation Science*, **2**, 307–14.

Trabaud, L. and Lepart, J. (1980) Diversity and stability in garrigue ecosystems after fire. *Vegetatio*, **43**, 49–57.

Trapnell, C.G. (1959) Ecological results of woodland burning experiments in northern Rhodesia. *Journal of Ecology*, **47**, 129–68.

Trollope, W.S.W. (1978) Fire behaviour – a preliminary study. *Proceedings of the Grassland Society of South Africa*, **13**, 123–8.

Trollope, W.S.W. (1982) Ecological effects of fire in South African savannas, in *Ecology of Tropical Savannas*, (eds B.J. Huntley and B.H. Walker), Ecological Studies 42, Springer-Verlag, Berlin, pp. 292–306.

Trollope, W.S.W. (1984) Fire in savanna, in *Ecological Effects of Fire in South African Ecosystems*, (eds P. de V. Booysen and N.M. Tainton), Springer Verlag, Berlin, pp. 199–218.

Trollope, W.S.W. (1993) Fire regime of the Kruger National Park for the period 1980–1992. *Koedoe*, **36**, 45–52.

Troumbis, A.S. and Trabaud, L. (1989) Some questions about flammability in fire ecology. *Acta Oecologica–Oecologia Plantarum*, **10**, 167–75.

Uhl, C. and Kauffman, J.B. (1990) Deforestation, fire susceptibility, and potential tree responses to fire in the eastern Amazon. *Ecology*, **71**, 437–49.

Underwood, R.J. and Christensen, P.E.S. (1981) *Forest Fire Management in Western Australia*. Forests Department of Western Australia, Perth.

Usher, M.B. (1972) Developments in the Leslie matrix model, in *Mathematical Models in Ecology*, (ed. J.N.R. Jeffers), Symposium of the British Ecological Society, 12, Blackwell, Oxford, pp. 29–60.

van Hensbergen, H.J., Botha, S.A., Forsyth, G.G. and Le Maitre, D.C. (1992) Do small mammals govern vegetation recovery after fire in fynbos? in *Fire in South African Mountain Fynbos*, (eds B.W. van Wilgen, D.M. Richardson, F.J. Kruger and H.J. van Hensbergen), Ecological Studies 93, Springer-Verlag, Berlin, pp. 182–202.

van Rensburg, H.J. (1972) Fire: its effects on grasslands, including swamps – southern, central and eastern Africa. *Proceedings of the Tall Timbers Fire Ecology Conference*, **11**, 147–73.

van Wagner, C.E. (1977) Conditions for the start and spread of crown fire. *Canadian Journal of Forestry Research*, **7**, 23–34.

van Wagner, C.E. (1978) Age-class distribution and the forest fire cycle. *Canadian Journal of Forestry Research*, **8**, 220–7.

van Wagner, C.E. and Methven, I.R. (1978) Discussion: two recent articles on fire ecology. *Canadian Journal of Forestry Research*, **8**, 491–2.

van Wilgen, B.W. (1981) An analysis of fires and associated weather factors in mountain fynbos areas of the south, western cape. *South African Forestry Journal*, **119**, 29–34.

van Wilgen, B.W. and Forsyth, G.G. (1992) Regeneration strategies in fynbos plants and their influence on the stability of community boundaries after fire, in *Fire in South African Mountain Fynbos*, (eds B.W. van Wilgen, D.M. Richardson, F.J. Kruger and H.J. van Hensbergen), Ecological Studies 93, Springer-Verlag, Berlin, pp. 81–107.

van Wilgen, B.W. and van Hensbergen, H.J. (1992) Fuel properties of vegetation in Swartboskloof, in *Fire in South African Mountain Fynbos: Species, Community and Ecosystem Response in Swartboskloof*, (eds B.W. van Wilgen, D.M. Richardson, F.J. Kruger, and H.J. van Hensbergen), Ecological Studies 93, Springer-Verlag, Berlin, pp. 37–53.

van Wilgen, B.W. and Viviers, M. (1985) The effect of season of fire on serotinous Proteaceae in the Western Cape and the implications for fynbos management. *South African Forestry Journal*, **133**, 49–53.

van Wilgen, B.W., Higgins, K.B. and Bellstedt, D.U. (1990a) The role of vegetation structure and fuel chemistry in excluding fire from forest patches in the fire-prone fynbos shrublands of South Africa. *Journal of Ecology*, **78**, 210–22.

van Wilgen, B.W., Everson, C.S. and Trollope, W.S.W. (1990b) Fire management in Southern Africa: some examples of current objectives, practices, and problems, in *Fire and Tropical Biota* (ed. J.G. Goldammer), *Ecological Studies* 84, Springer-Verlag, Berlin, pp. 179–215.

van Wilgen, B.W., Bond, W.J. and Richardson, D.M. (1992) Ecosystem management, in *The Ecology of Fynbos: Nutrients, Fire and Diversity*, (ed. R.M. Cowling), Oxford University Press, Cape Town, pp. 345–71.

van Wilgen, B.W., Richardson, D.M. and Seydack, A.W.H. (1994) Managing fynbos for biodiversity: constraints and options in a fire-prone environment. *South African Journal of Science*, **90**, 322–9.

Vasconcelos, M.J. and Guertin, D.P. (1992) FIREMAP – simulation of fire growth with a geographic information system. *International Journal of Wildland Fire*, **2**, 87–98.

Vasquez, A. and Moreno, J.M. (1993) Sensitivity of fire occurrence to meteorological variables in Mediterranean and Atlantic areas of Spain. *Landscape and Urban Planning*, **24**, 129–42.

Veblen, T.T., Hadley, K.S., Nel, E.M. *et al.* (1994) Disturbance regime and disturbance interactions in a Rocky Mountain subalpine forest. *Journal of Ecology*, **82**, 125–35.

Vines, R.G. (1980) Physics and chemistry of rural fires, in *Fire and the Australian Biota*, (eds A.M. Gill, R.H. Groves and I.R. Noble), Australian Academy of Science, Canberra, pp. 129–50.

Vogl, R.J., Armstrong, W.P., White, K.L. and Cole, K.L. (1977) The closed cone pines and cypresses, in *Terrestrial Vegetation of California*, (eds M.G. Barbour and J. Major), Wiley, New York, pp. 295–358.

Walker, B.H. and Noy-Meir, I. (1982) Aspects of the stability and resilience of savanna ecosystems, in *Ecology of Tropical Savannas*, (eds B.J. Huntley and B.H. Walker), Ecological Studies 42, Springer-Verlag, Berlin, pp. 556–609.

Walker, J. (1981) Fuel dynamics in Australian vegetation, in *Fire and the Australian Biota*, (eds A.M. Gill, R.H. Groves and I.R. Noble), Australian Academy of Science, Canberra, pp. 101–28.

Walker, J. and Peet, R.K. (1983) Composition and species diversity of pine-wiregrass savannas of the GreenSwamp, North Carolina. *Vegetatio*, **55**, 163–79.

Walker, J., Davis, J.R. and Gill, A.M. (eds) (1985) Towards an expert system for fire management at Kakadu National Park. *CSIRO, Division of Water and Land Resources Technical Memo 85/2*, CSIRO, Canberra.

Waller, D.M. (1979) Models of mast fruiting in trees. *Journal of Theoretical Biology*, **80**, 223–32.

Walter, H. (1971) *Ecology of Tropical and Sub-tropical Vegetation*, Oliver and Boyd, Edinburgh.

Watt, A.S. (1947) Pattern and process in the plant community. *Journal of Ecology*, **35**, 1–22.

Watt, A.S. (1955) Bracken versus heather: a study in plant sociology. *Journal of Ecology*, **43**, 490–506.

Watt, A.S. (1964) The community and the individual. *Journal of Ecology*, **52**, (Suppl.), 203–11.

Webb, N.R. (1980) The Dorset heathlands: present status and conservation. *Bulletin d'Ecologie*, **11**, 659–64.

Wedin, D. and Tilman, D. (1993) Competition among grasses along nitrogen gradient: initial conditions and mechanisms of competition. *Ecological Monographs*, **63**, 199–229.

Wein, R.W. (1983) Fire behaviour and ecological effects in organic terrain, in *The Role of Fire in Northern Circumpolar Ecosystems*, (eds R.W. Wein and D.A. MacLean), SCOPE, Wiley, New York, pp. 81–95.

Wein, R.W. and MacLean, D.A. (1983) An overview of fire in northern ecosystems, in *The Role of Fire in Northern Circumpolar Ecosystems*, (eds R.W. Wein and D.A. MacLean), SCOPE, Wiley, New York, pp. 1–20.

Weiner, J. (1990) Asymmetric competition in plant populations. *Trends in Ecology and Evolution*, **5**, 360–4.

Weller, D.E. (1987) A reevaluation of the –3/2 power rule of plant self-thinning. *Ecological Monographs*, **57**, 23–43.

Wellington, A.B., Polach, H.A. and Noble, I.R. (1979) Radiocarbon dating of lignotubers from mallee forms of *Eucalyptus*. *Search*, **10**, 282–3.

Wells, P.V. (1965) Scarp woodlands, transported grassland soils and the concept of grassland climate in the Great Plains region. *Science*, **148**, 246–9.

Wells, P.V. (1969) The relation between mode of reproduction and extent of speciation in woody genera of the California chaparral. *Evolution*, **23**, 264–7.

Wells, P.V. (1970) Post-glacial vegetational history of the Great Plains, new evidence reopens the question of the origin of treeless grasslands. *Science*, **167**, 1574–82.

Went, F.W., Juhren, G. and Juhren, M.C. (1952) Fire and biotic factors effecting germination. *Ecology*, **33**, 351–63.

Werner, P.A. and Platt, W.J. (1976) Ecological relationships among co-occurring goldenrods (*Solidago*: Compositae). *American Naturalist*, **110**, 959–71.

Westfall, R.H., Everson, C.S. and Everson, T.M. (1983) The vegetation of the protected plots at Thabamhlope Research Station. *South African Journal of Botany*, **2**, 15–25.

Westoby, M., Walker, B. and Noy-Meir, I. (1989) Opportunistic management for rangelands not at equilibrium. *Journal of Range Management*, **42**, 266–74.

Westoby, M., Jurado, E. and Leishman, M. (1992) Comparative evolutionary ecology of seed size. *Trends in Ecology and Evolution*, **7**, 368–72.

White, A.S. (1983) The effects of thirteen years of annual burning on a *Quercus ellipsoidalis* community in Minnesota. *Ecology*, **64**, 1081–5.

White, F. (1976) The underground forests of Africa: a preliminary review. *Gardens Bulletin, Singapore*, **29**, 57–71.

White, F. (1983) *The Vegetation of Africa*, UNESCO, Paris.

Whitney, G.G. (1986) Relation of Michigan's presettlement pine forests to substrate and disturbance history. *Ecology*, **67**, 1548–59.

Whittaker, E. and Gimingham, C.H. (1962) The effects of fire on regeneration of

Calluna vulgaris (L.) Hull from seed. *Journal of Ecology*, **50**, 815–22.
Wicklow, D.T. (1977) Germination response in *Emmenanthe penduliflora* (Hydrophyllaceae). *Ecology*, **58**, 201–5.
Williams, I.J.M. (1972) A revision of the genus *Leucadendron* (Proteaceae). *Contributions of the Bolus Herbarium, 3.*
Williamson, G.B. and Black, E.M. (1981) High temperature of forest fires under pines as selective advantage over oaks. *Nature*, **293**, 643–4.
Willson, M.F. (1983) *Plant Reproductive Ecology*, Wiley, New York.
Woodward, F.I. (1987) *Climate and Plant Distribution*, Cambridge University Press, Cambridge.
Woodward, F.I. (1992) A review of the effects of climate on vegetation: ranges, competition, and composition, in *Global Warming and Biological Diversity*, (eds R.L. Peters and T.E. Lovejoy), Yale University Press, pp. 105–23.
Wright, H.A. (1971) Why squirreltail is more tolerant to burning than needle-and-thread. *Journal of Range Management*, **24**, 277–84.
Wright, H.A. and Bailey, A.W. (1982) *Fire ecology: United States and Southern Canada*, Wiley, New York.
Wright, H.A., Bunting, S.C. and Neuenschwander, L.F. (1976) Effect of fire on honey mesquite. *Journal of Range Management*, **29**, 467–71.
Yeaton, R.I. (1988) Porcupines, fires and the dynamics of the tree layer of the *Burkea africana* savanna. *Journal of Ecology*, **76**, 1017–29.
Yeaton, R.I. and Bond, W.J. (1991) Competition between two shrub species: dispersal differences and fire promote co-existence. *American Naturalist*, **138**, 328–41.
Yoda, K., Kira, T., Ogawa, H. and Hozumi, K. (1963) Self thinning in overcrowded pure stands under cultivated and natural conditions. *Journal of Biology, Osaka City University*, **14**, 107–29.
Zammit, C.A. (1988) Dynamics of resprouting in the lignotuberous shrub *Banksia oblongifolia. Australian Journal of Ecology*, **13**, 311–20.
Zammit, C.A. and Westoby, M. (1987) Seedling recruitment strategies in obligate-seeding and resprouting *Banksia* shrubs. *Ecology*, **68**, 1984–92.
Zammit, C.A. and Zedler, P.H. (1988) The influence of dominant shrubs, fire, and time since fire on soil seed banks in mixed chaparral. *Vegetatio*, **75**, 175–87.
Zedler, P.H., Gautier, C.R. and McMaster, G.S. (1983) Vegetation change in response to extreme events: the effects of a short interval between fires in California chaparral and coastal scrub. *Ecology*, **64**, 809–18.

Species Index

Subject Index

GPSR Compliance

The European Union's (EU) General Product Safety Regulation (GPSR) is a set of rules that requires consumer products to be safe and our obligations to ensure this.

If you have any concerns about our products, you can contact us on ProductSafety@springernature.com

In case Publisher is established outside the EU, the EU authorized representative is:

Springer Nature Customer Service Center GmbH
Europaplatz 3
69115 Heidelberg, Germany

Batch number: 10809441

Printed by Printforce, the Netherlands